MODERN
WINEMAKING

Chardonnay grapes near harvest. Photo by Kenneth Garrett.

MODERN WINEMAKING

PHILIP JACKISCH

CORNELL UNIVERSITY PRESS

ITHACA AND LONDON

First published 1985 by Cornell University Press

Printed in the United States of America

Library of Congress Cataloging in Publication Data

Jackisch, Philip, 1935–
    Modern winemaking.

    Bibliography: p.
    Includes index.
    1. Wine and wine making—Amateurs' manuals.
I. Title.
TP548.2.J33   1985   663'.2   84-45803
ISBN 0-8014-1455-5 (alk. paper)

Illustrations by Nancy Bundorf

Cornell University Press strives to use environmentally responsible suppliers and materials to the fullest extent possible in the publishing of its books. Such materials include vegetable-based, low-VOC inks and acid-free papers that are recycled, totally chlorine-free, or partly composed of nonwood fibers.

Cloth printing   10

To Della, Margo, and Rita

CONTENTS

PREFACE

Winemaking has evolved over thousands of years but the pace of its development has increased greatly in the past few decades. It is now possible for anyone with access to grapes or other ingredients of decent quality to make consistently palatable or even excellent wines. This book describes techniques that have yielded good results for small winemakers, both amateur and professional. Because one of my premises is that all wines should be made to please consumers, information on evaluating and judging wines is included. Specific examples of methods and techniques are provided to bridge the gap between theory and practice.

Although this book is aimed largely at amateur winemakers, it assumes a professional attitude and contains information that small commercial winemakers can use. It should also interest sellers of winemaking supplies, wine instructors, and others who want to learn how wines are made and evaluated.

Most of the techniques in this book have been developed by others, but I have tried almost all of them out myself and have also exchanged information with hundreds of winemakers and wine consumers.

The first five chapters give an overview of the subject, the next four describe general winemaking practices, the following seven chapters deal with specific types of wines, and the last three concern using and evaluating finished wines. Appendixes give additional technical information and list some sources of supply.

As someone has said, borrowing heavily from one source is plagiarism but borrowing from many sources is scholarship. I have learned about winemaking from many books and through my association with members of the American Society of Enologists (especially the Eastern Section), the American Wine Society, the Home Wine and Beer Trade Association, Les Amis du Vin, the Society of Wine Educators, researchers from state universities and the United States Department of Agriculture, and grape growers and winemakers, both amateur and professional, in many states. I have also been greatly influenced by Leon D. Adams, Maynard A. Amerine, and Philip Wagner, three of this century's towering figures in American winemaking. Probably more than anyone else, they are responsible for the revival of good winemaking across the continent since 1933. All three are exceptional scholars and writers and have unselfishly shared their knowledge with winemakers everywhere.

My father, August, was the first amateur winemaker I ever met. He started making wines in Wisconsin before World War I. He and my mother, Della, steered me into science by buying me a chemistry set when I was ten years old and then put up with all my messy experiments. But the greatest influence has been exerted by my wife, Margaret, who insisted that she had a right to good wines even if they were made at home. As she became an expert and a wine judge, she raised her standards even higher. This steady prodding caused me to keep trying to find out how good wines can be made consistently. Every winemaker should be fortunate enough to have such a companion and consumer. Without her this book would not have been possible.

PHILIP JACKISCH

Baton Rouge, Louisiana

MODERN
WINEMAKING

1

An Overview of Winemaking

Winemaking predates recorded history, and writings thousand of years old suggest that appreciation of wine is as ancient as civilization. The enjoyment of wine has been one of the recurring high points in human experience.

The winemaking process can most easily be understood if it is divided into phases. First is the biological phase, during which grapes (or other raw materials) grow and ripen. At this time the basic quality of the wine is set; while winemakers can preserve this quality they can rarely improve it.

Next is the microbiological/enzymatic phase, called fermentation, when microorganisms (yeasts and bacteria) do most of their work. These microorganisms produce enzymes: yeast enzymes that convert grape sugars into alcohol and carbon dioxide; and bacterial enzymes that sometimes convert malic acid into the milder lactic acid. It is during this phase that the winemaker determines the basic style of the wine. One can, for example, make a low-alcohol, very fruity, semi-sweet white wine by conducting the fermentation at low temperatures and stopping it before it is finished. Or one can take the same grapes and make a higher-alcohol, complex, dry white wine with a touch of oak flavor by fermenting to dryness at higher temperatures in barrels.

The third phase is the physical or clarification stage, during which small particles in the wine settle by the force of gravity. At this time tartaric acid (a natural grape acid) combines with potassium (a mineral element found in grapes) and precipitates as potassium bitartrate (cream of tartar). Fining agents (special additives to be discussed later) combine with wine components, such as proteins or tannins, and settle out. Filtration removes visible particles and some or all of the remaining microbes. It is at this stage that the appearance and the stability of the wine are largely determined.

The fourth phase is the chemical or aging phase, during which the various components of the wine combine with oxygen or each other to form new substances. If the wine is stored in wooden barrels, some soluble extractives from the wood dissolve, adding odor and flavor. This is the time when the ultimate quality of many wines is realized.

These four phases occur in roughly the given order, but there is often some overlap. During grape ripening and prior to fermentation some enzymatic action takes place. During fermentation some settling of grape particles, yeast cells, and tartrates occurs. Aging begins during clarification. And during aging clarification can continue and sometimes unwanted yeast or bacterial activity takes place. While a winemaker cannot control all happenings in each phase, he or she should be aware of them and direct them toward the type of wine desired.

Winemaking Treatments

Grape juice is the basic material that the winemaker works with to produce the type of wine wanted. Things the winemaker can manipulate in or add to simple grape juice during winemaking include:

> Varietal character
> Extracts from dark grape skins
> Aging effects (and possibly wood extracts)
> Residual sugar
> Carbon dioxide
> Extra alcohol
> Oxidation products
> Special flavorings

Varietal character increases as grapes ripen and is generally greater in the cooler districts in which a given variety ripens. Winemakers usually favor high varietal character in wines derived from vinifera grapes and moderate or low varietal character in those made from native American or hybrid grapes. In many white wines, varietal character is enhanced by cool fermentations and in some instances special strains of yeast may modify it. Varietal character is lessened if grapes are harvested before they are fully ripe or if more neutral grapes or juice is blended in.

In making red wines a winemaker can control varietal character, color, tannins, and other substances derived from the grape skins by adjusting the time the skins remain in contact with the fermenting wine. Contact time varies from a few hours for rosé wines to days or weeks for heavier red wines.

The following example shows some of these steps.

EXAMPLE 1A

Making a simple red wine from grapes

To make 5 gallons of red wine, the winemaker selects 2 lugs or bushels (approximately 70 lb) of ripe, dark-skinned grapes and crushes them, by hand or with a mechanical crusher, removing as many stems as possible. The crushed grapes are placed in a 7–10-gallon crock or food-grade plastic container, and 2 grams of potassium metabisulfite dissolved in a little water are mixed into them. The crock is allowed to stand for several hours.

On the surface of the crushed grapes (called must), the winemaker sprinkles a packet of dried winemaking yeast. (Montrachet and Champagne yeasts are suitable and readily available.) After a few hours the yeast is mixed into the must. When fermentation starts, in a few hours or a day, the winemaker uses a potato masher to punch down the grape skins that have risen to the top, repeating this procedure several times a day. Between punchings, the container is covered with cheesecloth to keep out insects.

After 3 or 4 days, when fermentation is well underway, the juice is pressed from the grape pulp with a small wine press or other simple squeezing device and transferred to a 5-gallon glass carboy, which is stoppered with an air lock. The winemaker places any excess juice in smaller bottles filled nearly to the top and stoppered with air locks.

In 1–3 weeks fermentation should be completed—as evidenced by a lack of bubbles rising through the air lock. Using a plastic tube about ⅜ inches in diameter, the winemaker siphons the wine (a process called racking) into another carboy, taking care to leave the sediment (called lees) behind in the bottom of the first. The wine is racked from smaller containers in the same way and combined to fill the minimum number of containers. After 6–8 weeks it is racked again. In another 3 months it should be fairly clear. It is racked once more, then allowed to stand for about 6 months, during which the winemaker makes sure that containers are full and air locks filled with water.

At the end of the first year the wine should be quite clear and can be bottled. It is siphoned into gallon jugs, fifth-sized bottles, or other small containers that can be closed with a screw cap or cork. The wine should be ready to drink at this time but may improve with aging for a year or more.

When wines are aged, the aroma (which comes from the odorous components of the grapes) decreases, and the bouquet (which comes from newly formed chemical compounds) increases. Wines aged in wood casks pick up character from the wood. Red wines, which contain more tannins than whites, are usually aged longer to permit the bouquet to develop and the harshness from the tannins to decrease. As aging proceeds, the original fruitiness of both red and white wines diminishes; after long aging varietal character also diminishes.

Some wines are produced containing residual unfermented sugar. Sugar increases wine body, masks other flavors, and gives a sweet taste that novice wine drinkers often prefer. Experienced wine drinkers, on the other hand, usually prefer drier wines, in which other flavors are not masked by sugar.

Sparkling wines contain excess dissolved carbon dioxide, usually from a second fermentation in bottle or tank. This gas escapes slowly as bubbles when the sparkling wine is served and adds interest to its appearance, accentuates the fragrance, and provides an interesting tactile sensation in the mouth. Because dissolved carbon dioxide increases sourness, most sparkling wines contain some sugar to offset this effect.

During fermentation most yeasts stop working when the alcohol content exceeds about 14%. If distilled spirits are added to increase the alcohol level to 16–20%, the resulting wines generally resist further fermentation and bacterial action and are stable primarily because of the alcohol. Wines with an alcohol content above 14% (the upper legal limit for U.S. table wines) tend to have an undesirable alcoholic character unless something else is added to them. Port-type wines are made by stopping the initial fermentation by adding distilled spirits when a significant amount (about 10%) of sugar remains. The resulting wine is high in alcohol, sugar, and fruitiness and can withstand long aging periods. The highest-quality wines of this type, such as Portuguese ports, derive much of their quality from the original grapes.

When oxygen from the air reacts with wine, distinctive nongrape odors and flavors are produced. Sherry-type wines are made by adding distilled spirits to a fairly neutral white wine that is then allowed to oxidize slowly. Traditional Spanish sherries are produced by the action of a special surface yeast (flor yeast) or by prolonged storage in the presence of air. Sherry-type wines are also produced in the United States and elsewhere by a baking process involving months of heating. In all these methods, the original quality of the grapes is often less important than the processing used.

A variety of flavorings—sometimes fruit flavors—are added to wines to give distinctive types. Vermouth-type wines are made by adding distilled spirits to relatively flavorless white wines, then adding various herbs and spices and sometimes coloring matter.

Some wines are derived from fruits and are similar in properties and uses to white grape wines, both dry and sweet. Apples, pears, and stone fruits (peaches, apricots, and cherries) are most often the bases of commercial fruit wines, but other fruits and berries can also be used.

Thus, starting with grape or fruit juice, a winemaker can create an infinite variety of different wines. The influence of any modification depends on other factors present. For example, carbon dioxide can significantly affect the character of a dry Chardonnay wine, but may make less difference to that of a sweet, red Concord wine. If one considers only grape varieties, it is difficult to comprehend the differences and similarities of many European and American wines. Winemakers need to understand the tools they command in order to make classic styles of wines or interesting variations of them.

Craftsmanship and Artistry

Although winemaking has benefited greatly by the application of scientific principles (the major theme of this book), in many respects it remains a craft and an art. Like all craftsmen, winemakers need accurate information and an apprenticeship period, and need to pay careful attention to detail. In order to craft a wine, one must have goals in mind. Simply "to make the best possible wine" is too vague a goal and leads nowhere. A winemaker needs to formulate goals specifically in terms of: (1) available ingredients, (2) types of wine that will be well accepted, (3) available tools and techniques, and (4) the effort and cost that can be justified by the end result.

Wines can be works of art and most of the principles that apply to the production of other types of art also apply to great wines. Like other art forms, they can give pleasure and increase awareness. They can be distillations of experience. A sparkling wine, for example, recreates the magic of the fermentation process in a way that eliminates the unpleasant cloudiness and smells of the fermenting wine. It extracts and condenses one aspect of fermentation—the carbon dioxide gas that is produced along with alcohol—and it presents it to the wine consumer in an interesting and stimulating way.

One major value of works of art, including wines, is that they provide carefully controlled doses of sensory stimulation. No one form or quantity of stimulation is best in all circumstances. We want less when we are tired than when we are fresh. Some of us require different amounts and types of stimulation from others, because of both inherited requirements and conditioning. When various forms of stimulation are mixed, the result is frequently different from when they are presented separately. Most important, learning changes the type of sensory stimulation that each of us requires for optimum enjoyment. Different wines suit different people and circumstances. Moreover, the more complex a painting or a wine is, the more concentration is required to enjoy it fully. The most fabulous wines, therefore, have relatively limited uses. Even when we learn to appreciate the most interesting and expensive wines, all of us sometimes prefer simpler wines requiring less attention. Just as there is a place for background music, there is a place for background wines.

The Importance of Consumer Preferences

The place to start the winemaking process is with an understanding of what consumers prefer. Amateur winemakers should please not only themselves, but also family members and friends. They should begin by making a list of the wines that these people enjoy most during the year. Small commercial winemakers can check with local retail shops and restaurants to find out what is selling best in their areas, or what would sell if available at the right price. Winemakers should then think backward from the types of wines that they know consumers will enjoy to the ingredients and techniques at their disposal. If these are inadequate, they should try to change them. It is a mistake for a winemaker to use the handiest starting materials, apply some general winemaking methods, and see what happens. What usually results is a wine that only its maker can enjoy.

2
Grapes

Grapes are the most important fruit crop grown in the world, exceeding in quantity all other fruits combined. Grapes have been harvested for winemaking since prehistoric times, and most of the world's wines—including all the great ones—are made from grapes.

One reason why grapes are such a popular fruit is their high sugar content. They also contain fruit acids, minerals, and other substances that make them nearly ideal substrates for yeast fermentations producing wines. Tannins from grape skins retard oxidation of wines, and grape flavors last longer than do the flavors of other fruits. Grapes provide raw materials for a wide range of wine types.

Grape Varieties and Clones

Grapes belong to the botanical genus *Vitis* which includes two subgenera: *Euvitis* (true grapes) and *Muscadinia*. The former are sometimes called "bunch grapes" because they grow in bunches. The *Muscadinia* varieties grow as separate berries.

The subgenus *Euvitis* is divided into fewer than 60 species. Essentially all of those important in winemaking originated in the northern hemisphere. By far the most important species is *Vitis vinifera,* which is native to the area of Asia Minor south of the Black and Caspian seas. This Old World grape is the basis of most wines made today.

The United States has more different native species than any other country. Two species of Muscadinia are native to the southeast (*V. rotundifolia* and *V. Munsoniana*), but most native American vines are Euvitis species. In the latter group the only one used much in wine-making today is *V. labrusca*. However, several native American species are parents of hybrid grape varieties used as wine grapes or grape rootstocks.

Within a cultivated species, such as *V. vinifera,* there are many varieties, such as Chardonnay or Cabernet Sauvignon. Each variety has distinctive vine and fruit characteristics. When grapes are propagated from seeds, the daughter vines may not be identical to the parents. Long ago, grape growers found that they could retain the qualities of the parent vines by rooting cuttings from mature vine wood. Offspring produced in this way are "clones," as identical to their parents as it is possible to be.

When vines are successively propagated from wood cuttings, two factors can modify the quality of the daughter vines: virus in the parent vine, which will be transmitted to daughters and weaken them, and an occasional genetic mutation. Sometimes a grape grower finds an individual vine that differs slightly from those surrounding it. Occasionally such a vine will be more fruitful or the fruit may yield a superior wine. Such differences are probably due to absence of virus or to a positive mutation. If cuttings from this one vine are propagated and the difference holds up in the daughter vines, the result is known as a "clonal selection," a subvariety that possesses some special attribute. Grape growers now frequently purchase specific clonal selections when they plant new vineyards.

Grape Genetics and Hybrids

All wild *Vitis* vines are either male or female. When it was discovered that vines could be propagated from cuttings, growers who took cuttings only from fruitful vines failed because such vines were invariably female. The presence of unproductive male vines was essential to a productive vineyard.

At some time during the history of grape culture, a spontaneous mutation of flower type occurred, resulting in a vine with flowers that could fertilize themselves. This mutation occurred long ago among

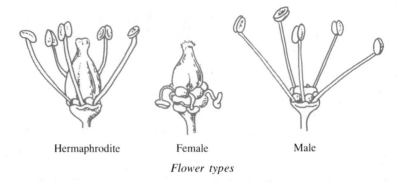

Hermaphrodite   Female   Male

Flower types

cultivated *V. vinifera* varieties, which now have "perfect flowers."
To cultivators, self-fruitful vines were a major advance. The wild
vines of North America do not have perfect flowers.

During the last century interest arose in producing new varieties by
deliberately introducing the pollen from one cultivated variety into
the flowers of another. European botanists imported American grape
cuttings to use, since the United States had many species—with the
unfortunate result that native American grapevine pests were inad-
vertently introduced into Europe. The first problem was the powdery
mildew fungus. It swept through European vineyards, where the vinifera vines had no natural resistance, until chemical sprays were developed to control it. Next to show up in Europe was the American
grape root louse, *Phylloxera vastatrix* (now known as *Daktulosphaira
vitifoliae*). This small insect is native to the eastern United States and
native vines possess roots too tough for it to destroy. But the roots of
European vinifera vines are more tender, and within a dozen years
following 1875 most of the important vineyards of France and central
Europe were devastated. The eventual solution to the problem was
found in the grafting of European vines onto rootstocks from American species with suitable resistance to the insect. For this purpose
American vines were imported into France in great numbers and
some, unfortunately, contained downy mildew and black rot.

In developing vines that would resist phylloxera, French botanists
crossed various American grape species with vinifera varieties. Some
of these new hybrids or crosses were able to grow on their own roots
and became known as "direct producers." Resistance to fungal dis-
eases was also sought through hybridization. Because these vines did
not require the expensive operation of grafting, they became popular
among the poorer French (and some other European) growers. In
many cases they yielded more grapes than the traditional *V. vinifera*

varieties. Higher yields contributed to excessive grape production in France, where for decades the government has attempted to limit hybrid plantings. Although the number of French-American hybrid vines in France is now decreasing, it is still greater than the total number of vines in the United States.

Whereas the California wine industry was founded on *V. vinifera* varieties, the eastern United States and Canadian wine industry was originally based on native American grapes, particularly *V. labrusca* varieties, which experience has shown possess more disease and weather resistance. It was long thought that varieties such as Concord and Delaware were pure labrusca. Recently, however, scientists have concluded that they are probably accidental crosses between native labrusca species (and possibly aestivalis species) and early trial plantings of vinifera varieties (which were made in the eastern United States from colonial times onward). Pure native American varieties give most unpalatable grapes and lack perfect flowers while commercial "labrusca" wines are reasonably palatable and the vines have perfect flowers. To distinguish these probable hybrids from the pure labrusca varieties, some grape scientists have proposed the term "labruscana" for them.

In the 1930s and 1940s grape growers in America (notably Philip Wagner in Maryland) discovered that imported French-American hybrids yielded grapes more suitable for winemaking than the native varieties. These hybrids, which were originally named after their developer with a number to indicate which of many crosses they were (e.g., Baco 1), have become the basis of a new wine industry in the eastern United States and Canada. As they became commercially important, new names were often given to them (e.g., Vignoles).

In Germany, California, and elsewhere, grape breeders have been busy producing crosses between vinifera varieties. In Germany, Muller-Thurgau has become the most widely planted variety. Many other crosses will undoubtedly be available in the future. Professor Harold Olmo of the University of California at Davis has worked to produce Euvitis-Muscadinia crosses, a project that could increase resistance to Pierce's disease. This disease (caused by a Rickettsia-like organism) wiped out vineyards in southern California in the 1880s and remains a serious problem in the hotter parts of California and most of the eastern United States south of the Ohio River.

Interspecific hybrids are sometimes a subject of controversy. The facts seem to indicate that essentially all commercial grape varieties are hybrids (accidental or deliberate) and that vinifera parentage confers palatability, while other parentage can confer resistance to dis-

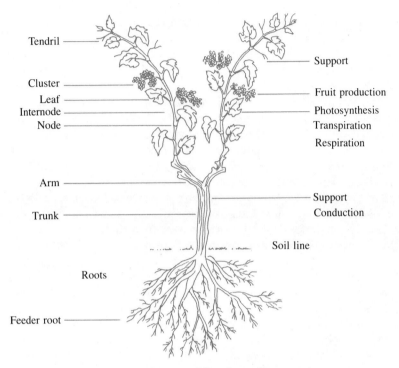

Tendril

Cluster
Leaf
Internode
Node

Arm

Trunk

Roots

Feeder root

Support

Fruit production
Photosynthesis
Transpiration
Respiration

Support
Conduction

Soil line

Important structures and functions of a grapevine

ease, insects, or hard winters. Where they can be successfully grown, the vinifera varieties are usually deemed preferable for the quality of their wines. But in some regions the various interspecific hybrids, with hardiness derived from American species, seem to have a place.

Factors That Influence Grape Quality

Modern winemakers can correct more grape deficiencies than their predecessors could, but no one can make a great table wine without using premium grapes. (Grape quality is less important for processed and flavored wines, such as sherries and vermouths.) Four factors have a significant effect on grape quality: climate, variety, cultural practices, and soil. Within a given climate, variety is the most important.

Weather is one of the greatest variables in winemaking. Vinifera varieties do best with rainy winters and long, warm, dry summers. It is optimal to have the grapes actually ripen during progressively cooling weather. Many of the world's greatest vineyard regions are located near large bodies of water, which moderate temperatures otherwise too extreme by retarding the warming of vineyards in the spring and their cooling in the autumn.

Grapes can be classified as ripening early, mid-season, and late, requiring, respectively, growing seasons of 130, 170, and 200 or more days. The best quality is achieved when a variety's ripening time closely matches the available growing season.

Most grapes require 20–30 inches of water each year. Excessive rain (or irrigation), especially during berry set or harvest, can greatly reduce crop size or quality.

Air temperatures from 77° to 86° F are optimal for growing most grapes. Leaf temperatures higher than 86° F can cause water deficiency, sunburn damage, and reduced growth rates. High temperatures reduce acids and increase the pH of wine grapes, inhibit color formation, and reduce normal aromas and flavors. Low relative humidity stresses grapevines. In the best European wine areas, with their varied soils, the amount of heat during the growing season appears critical; vintage years usually coincide with warm growing seasons. Sunshine may also influence quality. Grapes in northern regions receive more hours of sunshine during the middle of the summer than those grown further south. The sun's positive effect may be observed in Burgundy and the state of Washington, among other regions.

The question of which grape varieties are best remains open because the public's taste in wines changes. Certain vinifera varieties

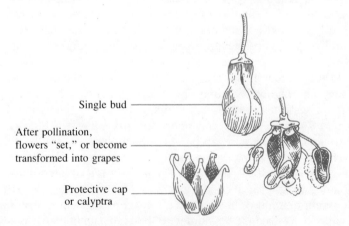

Single bud

After pollination,
flowers "set," or become
transformed into grapes

Protective cap
or calyptra

Stages in grape flower development

may be popular because they are versatile, capable of being made into wine of many styles. Chardonnay, for example, has been used to make light and fruity table wines, very full-bodied wines that can be aged in oak, great sparkling wines, and sweet wines with a botrytis character. Many of the highest-quality grapes tend to be "forgiving," so that even winemakers who are not exceptionally skilled can often make good wines from them. The highest-quality varieties give distinctive wines, interesting and complex without being too assertive or tiring on the palate. Some of these are memorable, a key factor in perceived wine quality.

The grape varieties used by the greatest number of wineries in the United States are generally those that have a worldwide reputation for quality. Popular among vinifera are Chardonnay, Cabernet Sauvignon, Zinfandel, Pinot noir, Riesling, Chenin blanc, Sauvignon blanc, and Gewurztraminer. Popular French-American hybrid varieties include Seyval blanc, DeChaunac, Aurore, Baco noir, Foch, Vidal blanc, and Chelois. The hybrids yield wines that are usually not as good as the best from vinifera varieties, are frequently too strongly flavored, and in many cases are probably best used in blends. Proponents, however, point out that in many eastern American locations these grapes give more consistently good results than the less hardy vinifera varieties. Popular native American varieties include Concord, Catawba, Niagara, and Delaware.

The category of cultural practices includes vine spacing, training vines on stakes or wires, and pruning to balance crop yield and grape quality. Vine spacing seems to have a relatively minor effect on yield and quality. Training systems are more important for some varieties, because they affect light exposure and sometimes air temperatures.

Experience has shown a roughly inverse relationship between crop size and grape quality. Overcropped grapes lose fruit character. The theoretical maximum yield per acre for any grapes is not much more than 15 tons. Some raisin grape varieties in California come close to this amount. In both Europe and North America, the most noble winegrape varieties, such as Chardonnay, Cabernet Sauvignon, Pinot noir, and Riesling, usually give poorer quality wines when producing more than 4 tons per acre. In less favorable climates, yields of 1 ton per acre may be the maximum for top wine quality. (One often sees claims of higher yields, but these are usually for isolated good seasons, not a long-term average.)

The main method of controlling crop size is the winter pruning of dormant vines to reduce the number of buds, which in theory controls

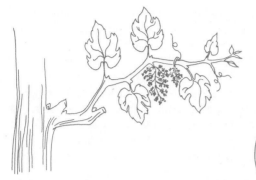

Early growth: A shoot with fruit clusters, tendrils, leaves, and a new bud at the base of each leaf

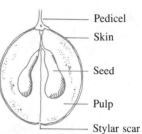

Pedicel

Skin

Seed

Pulp

Stylar scar

Longitudinal cross section of grape berry

grape clusters during the growing season to a level that experience has shown particular vines can properly ripen in an average year. In harsh climates, such as in the northeastern United States and parts of Canada, grape yields fluctuate widely from year to year, and pruning has only a partial influence on crop size.

Some winemakers prefer to have grapes pruned moderately and then have excess clusters or immature bunches removed during the growing season to gain a high foliage/grape ratio. But though this method may improve grape quality, it also increases vineyard expense, and relatively few winemakers feel it is justified. Cluster thinning is required with some of the French-American hybrid varieties and may also be needed with any type of vigorous virus-free grapevines if crop size has been underestimated at pruning time.

Winemakers operating vineyards can prune and otherwise manage grapevines as they wish. Winemakers purchasing grapes should realize that their goals may conflict with those of the grape growers. Most winemakers prefer small crops that stay on the vines until quality peaks, but growers can earn more from their vineyards if they grow a large crop and pick as early as possible before rots, birds, and frosts take a toll. In general, winemakers can get top quality in purchased grapes only by setting minimum quality requirements and being willing to pay growers enough to compensate them for their risks.

Commercial winemakers usually set minimum sugar levels for their purchased grapes. Grapes for dry table wines should be picked when their sugar content is just enough to give the desired alcohol content; overripe grapes may not ferment dry to produce an acceptable wine. Ripe wine grapes should have at least 0.7% acidity.

The influence of soil type on wine quality is disputed; soil fertility is usually less important than a soil structure favoring extensive root development. Vines with deep, well-spread roots start ripening grapes earlier and yield better-quality fruit. Grape crops are larger on very fertile soils, but the fruit is generally of lower quality. The highest-quality grapes, having the most intense flavor, usually come from shallow, not too fertile soils (often sandy or gravelly) with deep, well-drained subsoils.

The general California view is that soil type has less influence on quality than soil water content. Grapes grown in soils that retain water or grapes that are overirrigated will be less desirable, but in areas without sufficient summer rains, proper irrigation can help quality. Vines grown on bottom lands with excess moisture usually produce rather large crops of mediocre grapes. In California's north coast region, hillside vineyards are generally drier and produce a smaller and sometimes higher quality crop than valley-floor vineyards. Many quality vineyards in Burgundy and Bordeaux are on relatively level land, so slope itself is not a critical factor. Along the Rhine river, vineyards on eastern slopes receive more afternoon sunlight than those on level land, a factor that assists ripening in those northern latitudes.

There is evidence, at least in Europe, that when vines are under stress—from climate, soil, possibly even viruses—they produce better-quality grapes.

Because seasonal variations cause variations in quality, winemakers need to monitor vineyards to make their plans for the harvested grapes. The decision when to harvest is usually based on grape maturity, but condition may be a heavier factor if the grapes have been damaged by drought, rains, hail, frost, or rot. Winemakers' maturity standards should provide for a balance of alcohol, acids, and varietal flavors in the wine. Indicators of maturity include °Brix (sugar content), pH (the measure of acid vs alkaline content), and grape flavor.

Grapes smashed during picking are subject to oxidation before being delivered to the winery. Hand picking can usually avoid this problem, and grape shears are often preferred to knives since pickers have less tendency to squeeze grapes. Mechanical harvesting is now frequent in commercial vineyards. Grapes most suitable for mechanical harvesting have moderate amounts of foliage and firm berries that can be easily shaken from the vine. Quality can be maintained by the use of properly adjusted machines, field crushing, and the injection of sulfur dioxide (SO_2) into the must, which is then blanketed with carbon dioxide (CO_2). Mechanical harvesting provides advantages in cost and speed (an important factor before or just after bad weather), and properly done it can provide high-quality grapes for winemaking.

Purchasing Wine Grapes

Many winemakers, both amateur and commercial, begin by purchasing grapes, in one of the following forms: (1) fresh whole grapes, (2) machine-harvested grapes, (3) fresh juice, (4) frozen whole grapes, (5) frozen juice, and (6) concentrates. Each form offers certain advantages.

Fresh whole grapes, which permit the best control of the winemaking process, are usually preferred when the winemaker lives close

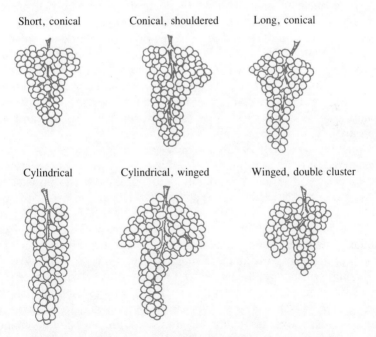

Short, conical   Conical, shouldered   Long, conical

Cylindrical   Cylindrical, winged   Winged, double cluster

Cluster shapes (the more open clusters resist rot better than the tightly packed. Cluster shape is largely dependent on grape variety)

enough to a vineyard to have them delivered soon after picking. With refrigeration, whole grapes can even survive a cross-country trip in a condition of reasonable quality. Some grapes are shipped in boxcars from California to eastern markets. Buyers have a limited choice of varieties and often can get only mediocre-quality grapes from California's hot Central Valley. Unrefrigerated grapes at a siding soon deteriorate and become moldy, but if one can get to a boxcar soon after its arrival and select the best-looking grapes, one can produce a palatable Zinfandel or other common wine.

In many cities one can arrange with a fruit retailer to bring in wine grapes—a more expensive option than buying at a boxcar siding, but the grapes will often be in better shape. Since fruit sellers are usually not knowledgeable about wine grapes, however, one may not get the grapes one ordered.

Those who live near commercial vineyards can sometimes find growers who will sell them grapes directly. Purchasers should, if possible, sample the grapes to make their own determination of ripeness and desired time of harvest.

Grape sugar content is usually determined with a hydrometer (which measures density) or a refractometer (which measures refractive index, a medium's bending of light). Before a hydrometer reading is taken, grape juice should be strained or filtered. A temperature correction should be made if that of the grapes varies from the calibration temperature of the hydrometer. A hand-held refractometer can measure the sugar content of individual grapes or a composite of many grapes, is not influenced by suspended grape pulp, and has automatic temperature compensation.

Amateur winemakers almost never purchase mechanically harvested grapes; many wineries do. This sort of harvesting creates problems because many grape skins are broken in the process. A discussion of the specialized equipment needed to handle these grapes properly is beyond the scope of this book.

Fresh juice from white grape varieties is sometimes available, from home winemaking supply shops and other sources. With proper handling (prompt crushing and pressing, SO_2 addition, prompt delivery), the quality of the grapes can be well preserved. The winemaker purchasing grape juice can avoid many handling problems but has no opportunity to separate underripe or moldy grapes before the pressing, no control over how hard the grapes are pressed, and no chance to verify that only grapes of the purchased variety were included in the pressing. Purchased grape juice is usually more expensive than juice produced by the winemaker.

The availability of frozen premium California grapes and juice across the United States and Canada gives many home winemakers an opportunity to work with some of the best varieties from some of the best vineyard regions. The cost of frozen grapes is high because of the special processing, handling, and shipping required, and the expense is not justified for grapes of ordinary quality. At present frozen grapes and juice are available only in 30-gallon and 55-gallon drums. The quality of the resulting wines is claimed to equal that of fresh grapes, but only home winemakers who can invest hundreds of dollars in grapes with the hope of producing outstanding wines will be interested in this source.

Concentrates are produced in many countries and cans or bottles of them are sold in most shops that handle home winemaking supplies. They are made by boiling grape juice at reduced pressure to eliminate about 80% of the water in it. In principle, concentrate is the easiest form of grapes for the novice to use, and many home winemakers begin this way. To make a wine, a user opens a can and adds water, possibly a package of additives, and yeast. Unfortunately, the wines that result are usually not of top quality. Many concentrates available to home winemakers are not made from the best grapes. Much quality, especially fragrance, is usually lost in the production process. Many concentrating operations use a high temperature which converts some sugar to unstable compounds, such as 2-hydroxymethyl-furfural, a substance with a caramel odor and taste that mar many amateur wines and can be instantly recognized by trained wine judges.

When red grapes are concentrated, the usual extraction of pigments and tannins that occurs during fermentation on the skins is not possible, even though some color and a little tannin is extracted by heating the grapes prior to pressing and concentration. Consequently, red grape concentrates make a paler, shorter-lived wine than the same fresh grapes would have made.

Concentrates deteriorate during storage, but producers have been lax about dating their products and providing adequate storage information to dealers. Ideally, concentrates should be refrigerated and sold within a few months, but most shops cannot afford refrigeration and nonetheless continue to sell concentrates for a year or more after receipt. During this time the concentrates darken and the unstable compounds in them polymerize to form hazes that are very difficult to remove from the resulting wines. Those sold in cans hold up better than those in plastic bottles, which permit oxygen from the air to diffuse into the concentrate and degrade it.

Despite their disadvantages, concentrates do have some role in making decent wines. Because they tend to give bland, low-acid wines, blending these with some of the overly assertive and high-acid wines found in much of the northeastern United States and Canada can often give a result superior to either ingredient alone. Good-quality rosé wine concentrates can often yield decent wines, equal to commercial rosé wines because many of the latter are weak in fragrance and varietal character and often have their shortcomings masked by a slight sweetness. Concentrates are also a good choice when a home winemaker wishes to produce a sherry, a vermouth, or other wine where the fragrance and flavor of the grapes is not important. The fact that concentrate contains less water than fresh grapes can be an advantage in achieving the body, high alcohol content, and sweetness found in most sherries.

Growing Wine Grapes

Many winemakers eventually get the urge to grow their own grapes. Benefits of operating a vineyard include healthful outdoor recreation and the pride of producing "estate bottled" wines.

General information on grape growing follows to help those considering it see what is involved. Books giving detailed directions are available, as is information from county agents and sometimes from state publications in areas where grapes are grown commercially on a significant scale.

Because the weather is uncertain in many growing districts, the grower is well-advised to plant several varieties with different ripening times, insurance against a total crop loss if the weather turns foul just before one harvest. A mixture of varieties is desirable also because many wines are better when they contain a blend of grapes. And of course most wine drinkers like some variety.

The potential grower needs an idea of how much land is required to produce the desired amount of grapes. Home and small commercial winemakers usually get about 150–160 gallons of juice per ton. Premium grapes yield about 1–4 tons per acre so wine production ranges from about 150 to 650 gallons per acre of vines.

In commercial American vineyards, the spacing between rows is often 10–12 feet, which allows room for normal farm tractors, and the spacing between vines in each row is 6–8 feet. On an average, there are about 600 vines per acre (giving each vine about 72 square feet).

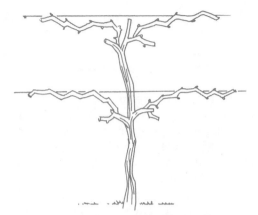

Four arm kniffen, a trellising system that with minor variations is popular in much of northern Europe and North America

Depending on the grape varieties, the climate, and other conditions, the average vineyard can produce between ¼ and 1 gallon of wine per vine. Even someone who has only an area 15 × 50 feet can grow a dozen vines and make up to 12 gallons (60 bottles) of wine per year. People with more room may consider giving a few acres to grapes and selling some of them.

Next, a potential grape grower needs to consider whether the land is of the right kind and in the right place. There are three things that grapevines cannot tolerate and continue to produce a crop: (1) low temperatures that kill the vines or buds, (2) shade, and (3) water-logged soils.

Two types of freezing affect grapevines. Severe winter weather can kill entire vines. Different varieties vary in winter hardiness, but most vinifera varieties show damage when temperatures drop much below 0° F. Many of the hybrid and native American varieties are more resistant. Snow provides insulation from short-term temperature dips, and with good snow cover, vinifera vines near the Great Lakes and the Finger Lakes in New York have survived winter temperatures down to −35° F. Cold drying winds can damage grapevines more than still cold, and a sudden drop in temperature after a warm spell can be very injurious. A second problem is "false springs" that freeze shoots and buds after they have started to grow. Some areas of the country that are free of severe winter lows are prone to false springs. Vinifera vines may grow all right where false springs occur, but seldom bear enough grapes to justify their planting.

Vinifera and other varieties require a winter rest period of at least two months, preferably with temperatures below 50° F. (This limitation probably is not as critical as was once thought since many vinifera varieties have been grown for a number of years in Louisiana and Mississippi.)

High humidity, a problem in some areas of the country, can foster the growth of various fungi and make it difficult to grow healthy grapes without an expensive spraying program.

The length of the growing season (from last spring frost to first autumn frost) largely determines what varieites of grapes can be successfully grown. In selecting a vineyard site, the grower should avoid frost pockets and areas lacking proper air drainage, which will effectively shorten the growing season at both the beginning and the end. Many amateur growers have a brick wall with southern exposure, which gives a few vines remarkable protection in severe climates.

Grapes can grow in almost any type of soil, but calcareous or slate soils favor the finest fruit production. Soil types that should be avoided include heavy clays; shallow, poorly drained soils; and soils containing above-average concentrations of alkali, boron, or other toxic substances. The physical texture of the soil is usually more important than its chemical composition. Ideally, grapes should be grown away from other crops. Not only shade, but the competition of other growth is undesirable. A large tree can reduce the fertility of a surprisingly large area of surrounding land. Fertilizers from nearby lawns or gardens can also affect grapevines and grapes. Too much nitrogen can lead to too much vegetative growth, while too much potassium can lead to excessively high pH levels in wines. Too much fertility is just as bad for wine grapes as too little.

The time required each year to care for one half-acre of grapes will be approximately:

Pruning vines	20 hours
Cultivation and weed sprays	10 hours
Planting, tying, and training vines	10 hours
Spraying vines for disease control	10 hours
Netting vines for bird control	4 hours
Harvesting and cleanup of vineyard	16 hours
Total	70 hours

The minimum tools and supplies needed to put in a new half-acre vineyard include:

Borrowed or rented tractor or cultivator
100 vineyard posts

Post-hole digger
Shovel
4000–6000 feet of trellis wire
Tools and supplies to string wire
300 rooted grapevines
Watering equipment for new grapevines

The minimum tools and supplies needed to care for an established one-half acre vineyard include:

Pruning shears
Vine ties and tying equipment
Borrowed or rented tractor or cultivator
Sprayer for herbicides
Sprayer for fungicides and insecticides
Nitrogen fertilizer (150 pounds of urea)
Herbicides, fungicides, insecticides
Grape netting (2000 × 14 feet)
Grape shears
Dozens of grape lugs or baskets

Actual amounts of nitrogen fertilizer needed will depend on soil fertility. Soil/petiole analyses (available through state agricultural agencies) are suggested to determine the optimum application of nitrogen. Mesural® is a chemical repellent that has proven effective in reducing bird damage and may completely eliminate the need for netting.

The economics of growing grapes on a small scale are difficult to generalize about. The price of suitable land varies from under $100 to well over $1000 per acre. Costs of bringing a vineyard into production also vary widely, depending on the condition of the land, the labor put in by the owner, and the local availability of vineyard supplies and help. The economic return from a vineyard operation depends primarily on crop size and quality. Depending on the growing region and year, the crop size from a half-acre vineyard can range from several tons down to a few hundred pounds, and grape quality can range from excellent to miserable.

People who live near their vineyards and average a ton or more of good-quality grapes per half-acre can probably justify the costs of growing their own. People in less favorable circumstances will have to decide if the effort is worthwhile from some standpoint other than the economic.

3

The Composition of Wines

Many compounds found in grapes are also found in wines, along with new compounds produced during fermentation or added later (in this chapter only grape wines will be considered). Typical ranges for these compounds as percents of the total are:

Compound	Grape must	Dry wine
Water	70–85	80–90
Carbohydrates (sugars)	15–25	0.1–0.3
Alcohols	0.0	8–15
Organic acids	0.3–1.5	0.3–1.1
Phenolic compounds	0.05–0.15	0.05–0.35
Nitrogen compounds	0.03–0.17	0.01–0.09
Carbonyl compounds	0.0	0.001–0.050
Inorganic compounds	0.3–0.5	0.15–0.40

Each of these classes of compounds (and others) will be discussed. Winemakers need to know what they are dealing with to make intelligent choices. Although some chemical formulas are given, readers need no knowledge of chemistry to use the information in this chapter.

Carbohydrates

The major soluble solids in grape musts and sweet wines are sugars. The sugar content of grapes is usually estimated from a hydrometer or refractometer measurement of total soluble solids, a valid procedure because more than 90% of the dissolved solids in grape juice are sugars. The units customarily used in the United States to express soluble solids are °Brix, where each degree is equal to 1 gram of sucrose per 100 grams of solution. The amount of soluble solids is a prime indicator of grape maturity.

Fermentable sugars are simple 6-carbon reducing sugars. Reducing sugars (named for their ability to reduce Fehling's solution) are those that contain one of the following reactive groups:

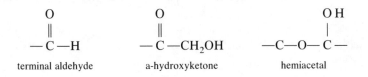

$$\underset{\text{terminal aldehyde}}{-\overset{\overset{\textstyle O}{\|}}{C}-H} \qquad \underset{\text{a-hydroxyketone}}{-\overset{\overset{\textstyle O}{\|}}{C}-CH_2OH} \qquad \underset{\text{hemiacetal}}{-\overset{\overset{\textstyle O\,H}{|}}{C}-O-C-}$$

The main sugars in grapes are glucose and fructose, both reducing sugars; the ratio of the former to the latter at grape maturity is about 1:1 but can range from 0.7:1 to 1.5:1. Fructose is approximately twice as sweet as glucose and about 1½ times as sweet as sucrose (table sugar). Small amounts of sucrose and other sugars are also present in grapes. When sucrose (a nonreducing sugar) is added to sugar-deficient grape musts, it is hydrolyzed to the two simpler sugars, glucose and fructose, aided by acids in the must or enzymes in the yeast.

$$
\begin{array}{cc}
\text{C HO} & \text{C H}_2\text{OH} \\
| & | \\
\text{H--C--OH} & \text{C==O} \\
| & | \\
\text{HO--C--H} & \text{HO--C--H} \\
| & | \\
\text{H--C--OH} & \text{H--C--OH} \\
| & | \\
\text{H--C--OH} & \text{H--C--OH} \\
| & | \\
\text{C H}_2\text{OH} & \text{C H}_2\text{OH} \\
\text{glucose} & \text{fructose}
\end{array}
$$

Dry wines contain less than 0.1% reducing sugar (often nonfermentable sugars such as pentoses). Detectable sweetness in wines varies with grape type and taster. In low-alcohol white wines the threshold for sweetness is about 0.4% but in some red wines it may be higher than 1.5%. Many table wines sold with reducing sugar levels above 0.5% find more acceptance by consumers than totally dry wines. The sugar content of a few wines is as high as 20 percent.

The riper grapes are, the more nonsugar solids (called "extract") will generally remain after fermentation, including organic acids, minerals, polysaccharides, proteins, and other substances. The extract of wines ranges from 0.7% to over 3.0%. The average value for white wines is about 2.0%, for reds about 2.5%.

The most important polysaccharides found in grapes are pectins, which consist of galacturonic acid and methyl galacturonate chains crosslinked with various sugars. The pectin content of grapes varies between 0.02 and 0.6% but is usually in the range of 0.1–0.2%; it is often higher in eastern grapes than in vinifera varieties. Gums and mucilages made of complex chains of sugars are also present in grapes.

Pectin-splitting enzymes are found in the skins of grapes, but when natural pectic enzymes are insufficient, winemakers often add commercial preparations. Pectins and gums exist in musts and in wines as negatively charged colloidal particles, tending to form hazes resistant to clarification. In some cases they act to prevent the precipitation of suspended materials such as ferric phosphate. Some winemakers believe that removing all of these colloidal particles by means of a tight filtration somewhat lowers wine quality, but pertinent sensory data is scarce.

Alcohols

All of the variety of alcohols found in wines contain a hydroxyl group (hydrogen and oxygen) attached to a carbon atom.

Ethanol, the principal alcohol in wines, is commonly called ethyl alcohol or, more informally, just alcohol. Throughout this book the common term "alcohol" will be used, except where confusion could result.

During fermentation, glucose and fructose are converted to ethanol and carbon dioxide (along with some side products). The overall conversion was first formulated by Gay-Lussac in 1815, and in modern chemical notation is:

$$C_6H_{12}O_6 \longrightarrow 2\,C_2H_5OH + 2\,CO_2$$

This equation predicts a theoretical yield of alcohol of 51.1% by weight or 59.0% by volume. The amount of ethanol actually formed depends on many factors including the amount of sugar in the grapes, yeast species, nutrient level of the grapes, and the temperature and general conditions of fermentation. Alcohol production will be lower than expected if the fermentation is incomplete. At higher temperatures less alcohol is found because more energy in the sugars is used by the yeast for growth and more alcohol is carried away by escaping carbon dioxide gas.

Various empirical formulas have been used to predict alcohol production during fermentation. For many grapes the amount of alcohol (in percent by volume) can be estimated to be 0.59 of the °Brix. For red grapes grown in warm areas a better estimate is 0.54 of the °Brix.

Yeasts or acetic bacteria in the presence of air can metabolize ethanol, especially in open red wine fermenters. Amateur winemakers, with much smaller fermenting vessels and much larger surface-to-volume ratios than commercial wineries, sometimes lose several percent alcohol in making red wines.

Alcohol in wines usually ranges from 7 to 14% (the U.S. government's limits for a table wine). In rare cases, with grapes very high in sugar and some semi-raisined grapes present, alcohol levels may reach 18% during fermentation. The desired alcohol level depends on a number of factors. In the absence of other ways to control bacterial growth, wines with less than 10% alcohol tend to spoil more easily than those above 10%. In California, commercial winemakers are prohibited from adding sugar to musts to achieve higher alcohol levels but can add alcohol to meet minimum alcohol requirements. California's former lower limit of 10% alcohol has now been changed, however, to allow the production of special light wines with alcohol as low as 7%. Modern sterile filtration and bottling technology have reduced the need for high alcohol levels to preserve wines.

Alcohol influences wine fragrance, taste, and body. A 4% (by volume) solution of alcohol in water is approximately as sweet as a 2% solution of glucose, and alcohol also enhances the apparent sweetness of sugar solutions. Alcohol moderates wine taste and usually makes it more pleasant. The amount of alcohol in red wines should generally be above 10%; at lower levels they tend to taste thin and bitter because tannins increase the threshold of sweetness detection and reduce apparent sweetness. On the other hand, wines having more than about 14% alcohol tend to have an alcoholic taste or fragrance sometimes described as "hot."

Dilute alcohol solutions in water have a greater viscosity than water alone so alcohol in wine contributes to a feeling of thickness on the tongue—to wine "body."

Other alcohols besides ethanol are found in wines. The simplest, methanol (CH_3OH), which contains one less carbon unit than ethanol (CH_3CH_2OH), is present in very small amounts in grape wines, but sometimes at a higher level in those made from other fruits. Most methanol in wines comes from the splitting of pectin by enzymes. This alcohol does not have a strong odor and contributes nothing to fragrance or flavor. The methanol content of grape table wines is generally less than 200 parts per million (ppm), far below toxic levels.

Longer chain alcohols, formed during fermentation and responsible for some of the complex fragrances of wines, include isopropyl alcohol, isobutyl alcohol, and several amyl alcohols—collectively called fusel oils. Their concentration in wines is about 50–200 ppm. Another alcohol found in some wines in small amounts is 2-phenylethanol. It has a smell reminiscent of roses and is most prevalent in the muscadine varieties native to the southeastern United States.

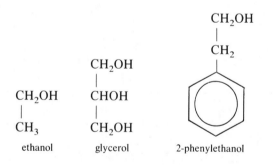

ethanol   glycerol   2-phenylethanol

Some wine alcohols are polyols or polyalcohols having more than one hydroxyl group per molecule; the major one is glycerol (commonly called glycerine). Normal table wines contain 0.5–1.5% by weight of glycerine, with an average of about 0.7%. Glycerine content depends on fermentation temperature, yeast strain, pH, initial sugar concentration, and aeration conditions. The addition of large amounts of sulfur dioxide increases glycerine production. Red wines usually have more glycerine than whites because they are fermented at higher temperatures. Though glycerine is sweet and very viscous, it is present in wines in such a small concentration that it seldom affects the taste or viscosity. The threshold for detecting glycerol in wines is about 1.5%.

2,3-Butanediol is another sweet-tasting polyol found in wines at about 500 ppm. Sorbitol, a sugar alcohol, is present in many fruits but its concentration in grapes is low. Another sugar alcohol, mannitol, is produced by lactic acid bacteria, and its presence in significant amounts signals bacterial spoilage. This reaction occurs in "stuck" fermentations, usually at high temperatures. *Botrytis cinerea* rot can also cause an increase in polyols such as mannitol.

Organic Acids

Organic acids contain a doubly bonded oxygen atom and a hydroxyl group, both attached to the same carbon atom. Acids give grape juice and wines their characteristic taste—ideally, clean and slightly tart. Other wine components—especially alcohol, sugars, and minerals—moderate the sour taste of acids and give balance. Adjusting acidity is an important task in winemaking.

Tartaric and malic are the major grape acids; tartaric acid and its salts normally provide more than half the total acidity of musts and wines. Because this acid is not respired, as malic acid is, during grape ripening, grapes in warm climates tend to have a higher ratio of it. The tartaric acid content of grapes varies from about 0.2 to 0.8% depending on variety, region, and temperature during the growing season. Tartaric acid is also the strongest acid present in grapes and wines, and it and its potassium salts largely control wine pH (effective acidity) which affects color, resistance to bacterial infections, and taste. Without tartaric acid, most wines should be deficient in these qualities.

As potassium moves into the berries, tartaric acid is partly converted from a free acid to salts, largely potassium bitartrate (cream of tartar); more than half of it in that form by harvest time. Since this salt is less soluble in alcohol than in water, during fermentation some of it precipitates.

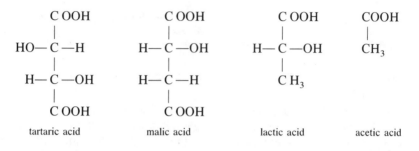

tartaric acid   malic acid   lactic acid   acetic acid

Malic acid is also found in many other fruits and vegetables. Its name comes from the Latin word for apple, and some white grape wines from cool growing regions, high in malic acid, sometimes have a taste reminiscent of apples. At high temperatures malic acid is respired; it may comprise only 10 to 40% of the total acid in warm climates but up to 70% in cool climates. It is sometimes reduced in wines by a malolactic fermentation, which converts it to the weaker lactic acid. Lactic acid is a minor byproduct of fermentation, and usually less than 0.1% is present in wines, but up to 0.6% is produced during a malolactic fermentation. Lactic acid has a mild sour taste and is found in buttermilk, sour cream, and cheeses. Certain undesirable lactic bacteria can infect wines and produce, along with lactic acid, smells reminiscent of spoiled milk or sauerkraut.

Citric acid is found in small amounts in grapes and wines. It is sometimes added by winemakers to increase acidity or to complex iron to prevent iron phosphate hazes.

Succinic acid, a minor grape acid, increases during the fermentation at about $\frac{1}{100}$ the rate of alcohol increase; thus about 0.1% succinic acid is found in wines.

Total acidity of must and wine is expressed in the United States as though all the acid were tartaric acid and is reported either as a percentage or in grams per liter. Acidity is measured by a procedure known as titration.

Most acids found in normal wines are nonvolatile and odorless, but acetic acid, found in small amounts in wines, is volatile. (It is the main acid in vinegar.) When wines containing certain bacteria are exposed to air, acetic acid is formed in large amounts along with ethyl acetate (the ester of acetic acid and ethanol). Much of the smell of wines turning to vinegar is caused by ethyl acetate, but analytical methods can more easily detect acetic acid so measurements of "volatile acidity" are used to monitor acetification. When acetic acid exceeds about 0.1%, most consumers can detect a vinegar smell in a wine and it is generally considered spoiled. All major winemaking countries set limits—generally in the range of 0.10 to 0.15%—on the amount of acetic acid permitted in sound wines.

Grape juice should ideally have about 0.7–0.9% acidity and table wines should have slightly less (0.6–0.85%, depending on sweetness). In the eastern United States, where grapes often have natural acidities above 0.9%, federal regulations permit the addition of water to decrease acidity to a lower limit of 0.5% (providing that no more than ⅓ water is added to the wine). Dessert wines generally have 0.4 to 0.65% acidity.

Phenolic Compounds

Phenolic compounds have hydroxyl groups (as do alcohols) but because these are attached to an aromatic ring, phenol properties differ from those of alcohols. The chemical structures of two phenolic compounds are shown below.

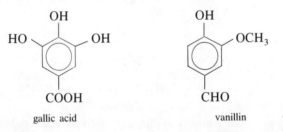

gallic acid   vanillin

Although the various phenolic compounds, such as pigments, in musts and wines are present in small amounts (usually below 0.05% in white wines and 0.3% in reds), they are among the most important constituents. These compounds are responsible for color, bitterness, astringency, some odors, some flavors, antioxidant activity, browning, and (in old red wines) dark precipitates. They come principally from grape skins, grape seeds, and oak barrels used in aging. The maximum average total phenolic content in red grapes is 5500 ppm, in white grapes 4000 ppm, and in stems 2000 ppm. About ¼–⅓ of the phenols are in the skins, and most of the rest in the seeds. Factors affecting wine's phenolic content include grape variety, total phenols in the grapes, skin and seed contact time, alcohol concentration, fermentation temperature, agitation of juice and skins, and intensity of pressing. In California, white wines average 250 ppm total phenols and reds about 1400 ppm. (It is common practice to express total phenols as though they were all gallic acid.)

Anthocyanin pigments, responsible for the brilliant red and blue colors of flowers and fruits, are the only significant pigments in red grapes. Anthocyanins constitute 100–500 ppm in young red wines, which means they have a small, approximately threshold effect on the flavor. But free anthocyanins disappear as wines age and pigments condense and precipitate. They can also be discolored by sulfur dioxide or by raising the pH.

A major anthocyanin pigment in vinifera varieties is malvidin with one glucose unit attached, called malvidin glucoside. Interspecific hybrids usually have two glucose units and are diglucosides, a difference that has been used to differentiate vinifera wines analytically from hybrids.

Other flavonoid phenols found in wine include anthocyanogens, catechins, flavonols (including light-yellow anthoxanthin pigments), and flavanones. Flavonoids provide flavor in red wines (but are essentially absent from whites, because skin contact is minimized) and are usually present at 5–10 times threshold taste levels. These compounds are found mostly in grape skins, except for the flavanones, which are found mainly in seeds.

Catechins, a large portion of total wine phenols (up to 100–200 ppm for whites and 1000 ppm for reds), are bitter but not astringent. The intermediate and higher tannins (polymers of catechins) are astringent but less bitter. Tannins, which receive their name from their ability to render gelatin insoluble, and thus tan hides to leather, are divided into the hydrolyzable (esters of gallic or ellagic acid) and the non-hydrolyzable. Their ability to combine with proteins makes it possible to clarify (''fine'') red wines with gelatin or egg whites. During fining the insoluble tannin/protein complex settles out, carrying additional particles with it. (Other bitter phenolic compounds, such as gallic acid, are also removed by fining.)

Flavanols and flavanones are very minor constituents of white wines, but reds can have 20–100 ppm. Leucoanthocyanins, found mostly in seeds, provide astringency and aid in fining.

Pigments and tannins, though not very soluble in cold water, are more so in hot water and in alcohol solutions. Early in a red wine fermentation tannins are precipitated by grape proteins or yeast; the major phenolic compounds in solution at this stage are the pigments. But once the proteins are gone, tannins enter and remain in the wine.

In the commercial production of eastern United States wines from grapes such as Concord, musts are often heated to increase the amount of color extracted from the skins. This practice, called ''hot pressing'' or ''thermal vinification,'' dissolves fewer tannins than a normal fermentation with skin contact does, but since these wines are rarely aged, lack of tannin is usually not a disadvantage.

Another class of wine phenolics is the nonflavonoid phenols, which provide no flavor, but some fragrance. Nonflavonoids can be volatile (e.g., tyrosol, syringaldehyde, and vanillin) or nonvolatile (e.g., gallic acid). They are largely derived from the grape juice but can also come from oak barrels and from the action of molds, bacteria, and yeasts. Their being both present in the juice and formed in fermentation means that their level is less sensitive to processing conditions than that of other phenols. Volatile phenols are usually present below odor threshold levels (1–50 ppm), but may contribute to odor through additive or synergistic effects.

A number of hydrolyzable tannins are found in wines exposed to oak barrels, oak chips, or oak extracts. The taste threshold for these extractable phenols is about 7–15 ppm.

Nitrogen Compounds

A wide variety of compounds containing nitrogen are found in wines, including ammonia, nitrates, amines, amino acids, peptides, proteins, and vitamins. They are important chiefly because they stimulate yeast and bacteria growth. The average total nitrogen content of grape musts is about 600 ppm. Chemical formulas for two compounds are given below.

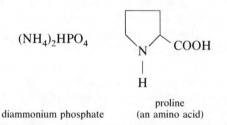

$(NH_4)_2HPO_4$

diammonium phosphate

proline
(an amino acid)

Ammonia appears in musts and wines largely as the ammonium ion (NH_4^+). In musts it ranges from about 5 to 175 ppm, averaging about 125 ppm. The nitrogen content of soils and its form greatly influence the ammonium content of musts. Yeasts use up much of it in fermentation; ammonia concentration in wines is only about 12 ppm.

For wine made with fruits and other nongrape ingredients, adding nitrogen—typically diammonium phosphate or urea—has been common. In the past most winemakers felt that grape musts contained enough nitrogen, but current evidence suggests that adding ammonium salts to them can improve wine quality. On the other hand, added ammonia can increase the amount of certain amino acids, especially histidine (a precursor of histamine), which may in some cases be undesirable.

Many amino acids are present in musts and wines, the principal one usually being proline, at an average concentration of about 500 ppm. Normal wine yeasts can make needed amino acids from ammonium ions and the sugar in musts, but those already present in grape musts can stimulate the growth of yeasts and increase the fermentation rate. During yeast growth many of the amino acids decrease. Some are necessary for the development of lactic acid bacteria.

Several physiologically active amines are found in wines, apparently formed by malolactic and other lactic bacteria. Among them are tyramine, which causes an increase in blood pressure, and histamine, which does the opposite. For most wine drinkers these effects are of negligible importance, but a few people experience stuffy noses or other allergy symptoms when consuming some red wines, a reaction blamed on histamine.

Proteins (complex chains of many amino acids) are found in grapes and some remain in wines, up to several hundred ppm. They precipitate during both fermentation and aging. Precipitation is increased at certain pH values and can sometimes also be caused by the blending of two wines.

A protein haze shows up in some white wines, a serious problem if it appears after bottling. Because in many cases it is related to high storage temperatures, the problem is often called "heat instability." The haze may result from a complex of protein with phenolic compounds. The protein content of wine can be reduced by fining with bentonite, heating grape musts, or (for white wines) adding small amounts of tannin.

Other nitrogen-containing components in musts and wines, such as nitrates and nitrites, a variety of amines, vitamins, nucleotides, and peptides (short chains of 2–4 amino acids) are found only in small amounts, and are of little importance to the winemaking process and wine character.

Carbonyl Compounds

Carbonyl compounds (aldehydes and ketones) are formed during fermentation or under oxidizing conditions. Aldehydes contain a doubly bonded oxygen atom and a hydrogen atom attached to a carbon atom. Though many are found in wines, only a few are present in much concentration or are of much significance. Volatile aldehydes usually have a pungent aroma. The chemical structures of two aldehydes sometimes found in wines are:

$$CH_3CHO \qquad HOCH_2 \diagdown_{O}\diagup CHO$$

acetaldehyde   hydroxymethylfurfural

Acetaldehyde is an intermediate in alcohol production but during fermentation it is almost all reduced to ethanol. In table wines its presence in amounts above 50 ppm signals unwanted oxidation, but in oxidized wines, such as flor sherries, a concentration of over 300 ppm is considered desirable. The detection threshold of acetaldehyde in water is about 1.5 ppm, but in wines, where sulfur dioxide is present, it may be 100 ppm. Acetaldehyde's odor depends on its concentration; at moderate levels it has a somewhat nutty smell.

Acetaldehyde is thought to be partly responsible for so-called "bottle sickness," a "faded" odor found in recently bottled wines that disappears after a few weeks. A faded odor also occurs during other stages of wine processing. Sulfur dioxide dissolved in water reacts with acetaldehyde to form a nonvolatile bisulfite complex. This removes the smell of acetaldehyde and gives wine a fresher fragrance. The reaction is reversible, however, and if sulfur dioxide is depleted by volatilization or oxidation, the acetaldehyde smell can return.

This same bisulfite reaction occurs during fermentation and a high level of sulfur dioxide traps acetaldehyde and prevents its reduction to ethanol. The result is more acetaldehyde at the end of fermentation than when less sulfur dioxide is used.

Acetaldehyde reacts with pigments in red wines, so its concentration is usually lower in reds than in whites. The effect of the reaction is to intensify color, especially the violet element.

Hydroxymethylfurfural is an aldehyde formed by the dehydration of fructose, possible when grape juice is concentrated at too high a temperature. Up to about 300 ppm is found in sherries and other baked wines and in Malaga, which is made from partly raisined grapes. The odor of this chemical has been described as being caramel-like, and its presence mars many otherwise acceptable amateur wines made from concentrates.

Ketones, chemically similar to aldehydes but less reactive, contain a doubly bonded oxygen atom attached to a carbon atom. Diacetyl and acetoin, ketones found in wines produced with too much air oxidation, are also formed during malolactic fermentations and may greatly influence a wine's odor if some wild strains of lactic bacteria are involved. The concentration of acetoin found during normal fermentations is 25–100 ppm but later it decreases. The normal concentration of diacetyl in wine is about 0.2 ppm. Above about 0.9 ppm it can yield an odor like sour milk. Diacetyl is also found in heated butter, and some people detect a hot butter odor in certain wines that have undergone a malolactic fermentation.

Inorganic Compounds

Inorganic compounds come from soil, in contrast to organic compounds, produced by living things. Inorganic substances in wines generally equal about 10% of the sugar-free extract. Such compounds consist of a positively charged metal atom or the ammonium group (a "cation") and a negatively charged element or group (an "anion"). Common table salt is an inorganic compound consisting of a sodium cation (Na^+) and a chloride anion (Cl^-).

Typical inorganic cations and anions found in wines are shown in the following list:

Cations	Chemical formula	Typical conc. in wine
potassium	K^+	1000 ppm
sodium	Na^+	80
calcium	Ca^{2+}	50
magnesium	Mg^{2+}	100
iron	Fe^{2+} or Fe^{3+}	2
copper	Cu^+ or Cu^{2+}	0.15
Anions		
chloride	Cl^-	60
phosphate	PO_4^{3-}	300
sulfate	SO_4^{2-}	700

Potassium is the most important cation found in growing things. The potassium content of grapes depends on variety, soil, climate, time of harvest, and other variables. During fermentation much of the potassium in grape musts precipitates as rather insoluble potassium bitartrate. Fermentation and storage temperatures, among other factors, influence the potassium in finished wine. White table wines usually have less than reds.

Sodium is very common in soil but is not taken up by plants as much as potassium is. The natural sodium concentration in wines is about 35 ppm. Certain additives (e.g., sodium bisulfite) increase that figure. Some commercial wineries use ion exchange before bottling, exchanging sodium for potassium to preclude potassium bitartrate precipitation. Because this treatment significantly raises sodium levels, its use is declining.

Calcium is also very common in soil and some is naturally found in wines. It sometimes forms calcium tartrate or calcium oxalate precipitates, which are troublesome if they settle slowly or appear after the wine is bottled. Calcium content rises if a must or wine is treated with calcium sulfate (to increase acidity) or calcium carbonate (to decrease acidity).

Magnesium usually is not of much concern to the winemaker, but it may influence tartrate stability and the acid taste of wines.

The iron content of new wines that have not come in contact with iron or regular (not stainless) steel equipment during processing usually averages about 1–2 ppm. Much (25–80%) of the iron in grape musts is removed during fermentation, primarily by yeasts. When iron is present in concentrations greater than 7–10 ppm, it can cause cloudiness or increase wine oxidation. The state of oxidation of iron depends on that of the wine. In wines kept away from air, 80–95% of the iron is in the ferrous (Fe^{2+}) state, but ferric iron (Fe^{3+}) increases when wines are aerated. The precipitation of colloidal ferric phosphate is responsible for a cloudiness in wines known as "white casse." Citric acid protects against white casse by forming a soluble ferric citrate complex. Ferric iron can also react with polyphenol compounds and precipitate as a blue-black film, known as "blue casse," in red wines.

Musts and wines generally contain only about 0.1–0.3 ppm of copper. Bordeaux mixture (a combination of copper sulfate and lime) employed in the vineyard may introduce larger quantities of copper into musts, as can contact with alloys containing copper (brass, bronze). Copper cloudiness, which occurs in wines containing more than 0.2–0.4 ppm of copper, may be caused by a copper-protein complex.

The main anions found in wines are tartrate, malate, lactate, acetate, nitrate, chloride, phosphate, and sulfate. Some have already been discussed. Phosphate normally comes from the soil the grapes are grown in but may also derive from yeast nutrients added to musts. Small amounts of sulfate are present in normal musts. During fermentation some yeasts can reduce sulfate to sulfur dioxide or even hydrogen sulfide; those strains that produce a minimum of sulfide are preferred. In the production of flor sherries, calcium sulfate is sometimes added to increase acidity. The increased sulfate concentration can give a slightly bitter taste, which may not be objectionable in sherries.

Odorous Compounds

Odor requires volatility. Hundreds of volatile compounds that contribute to the aroma or bouquet of a wine have been identified. (By convention aroma refers to odorous compounds from grapes and bouquet to compounds produced during fermentation and aging. Together they make up a wine's fragrance.) Many of these are found in only trace amounts, sometimes below threshold levels, but by additive or synergistic effects even trace components can influence wine odor.

Volatile compounds in wines include alcohols, organic acids and their esters, phenolic compounds, and carbonyl compounds. The main one (other than water, which is odorless) is ethanol, which makes up about 90% of the molecules entering one's nose when one sniffs a wine. High ethanol concentrations tend to mask the odor of other components, but fortunately, ethanol has only a moderate odor so other odors can make themselves felt.

Fruity smells are often associated with esters, reaction products of organic acids and alcohols. The chemical structures of two important esters found in some wines are:

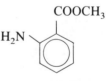

$CH_3COOCH_2CH_3$

ethyl acetate   methyl anthranilate

Although organic acids and alcohols can react directly to form esters, this process is very slow and limited in the presence of water. Most of the esters found in wines of normal age are either carried over from the grapes (or other fruit) or produced by yeast enzymes during fermentation. When yeasts stop growing, most ester formation stops.

Factors favoring ester formation include removal of suspended solids before fermentation, use of good wine yeast strains, maximum yeast cell growth, and high fermentation temperatures. But although more esters are formed at higher temperatures, they are also lost more rapidly by volatilization. A fermentation temperature of 54–59° F has been found to give the most esters in the final wine.

Total volatile esters in wines (calculated as ethyl acetate) average 200–400 ppm. The two most common are ethyl acetate (with a threshold of 160 ppm) and isoamyl acetate, both generally found at below threshold levels.

Odorous compounds in some grapes are so pungent that the wines have a very characteristic smell. Varieties with *V. labrusca* parentage contain 0.1–1 ppm of methyl anthranilate, giving wines with a unmistakable odor similar to that of Concord grape juice or grape-flavored pop. Some French-American hybrids and other varieties contain "leaf aldehydes" which impart a vegetal odor to wines. Muscat varieties (and to a lesser extent Riesling and some other varieties) contain alcohols called terpineols. One of these, linalool, has a smell very reminiscent of muscat wines. Cabernet Sauvignon contains a nitrogen compound also found in bell peppers, and some Cabernet Sauvignon wines have a green pepper smell.

Distinctive odorous substances sometimes decrease during oxidation and aging. Concord grapes can be made into sherry with little of the usual Concord odor. Old California Cabernet Sauvignon wine is sometimes difficult to distinguish from similarly aged Zinfandel.

Gases

Several gases dissolve in wines to some extent. The most important are oxygen, carbon dioxide, nitrogen, hydrogen sulfide, and sulfur dioxide (which is discussed as a chemical additive).

Oxygen does not directly contribute to fragrance or flavor, but the oxidation/reduction potential of must or wine depends mainly on the amount of dissolved oxygen. When oxygen is present, components of musts or wines can be oxidized; when it is absent, they can be reduced. Such reactions significantly affect wine aging and overall quality.

The approximate saturation concentration of oxygen in a wine is 8 ppm. During yeast growth, oxygen is readily used by the yeasts and removed. After yeast growth ceases, dissolved carbon dioxide protects the wine from excessive oxidation through the first racking or two. After that time oxygen is again absorbed by the wine and causes certain chemical changes by combining with other components. The oxygen level is influenced by the amount of sulfur dioxide, phenols (such as tannins), ascorbic acid (vitamin C), iron, copper, and other substances.

Oxygen dissolved in a must at the start of fermentation is desirable because it accelerates yeast growth and fermentation rate. During the aging of red wines some oxygen gets in, primarily during rackings and barrel toppings, perhaps as much as 1000 ppm. This oxygen ought to be introduced very slowly. After the wine has been slowly oxidized and certain phenolic compounds have polymerized, oxygen should be kept away from the wine. For most white wines it is best to keep oxygen contact to a minimum during cellar treatments and aging.

Carbon dioxide is formed in large amounts as fermentation proceeds. Its saturation concentration is generally governed by the alcohol level and the temperature and ranges from about 0.1 to 0.3%. It is also formed in sizable amounts during a malolactic fermentation. When this gas dissolves in wine, it combines with water to form the sour-tasting carbonic acid. The taste threshold for carbon dioxide in wines, where "spritz" can be detected, is 0.05–0.06%. For white wines, at a level between 0.06 and 0.12% it seems to increase the "freshness" of the aroma. In sparkling wines produced by bottle fermentation, dissolved carbon dioxide is one of the main quality factors. The dissolved gas may form loose complexes with proteins or other components and is then released more slowly when the wine is opened than it would be from wine that was merely carbonated. Autolysing yeast cells are probably a source of compounds that complex with carbon dioxide, and sparkling wines stored in contact with yeast for months or years show superior bubbles.

Nitrogen, the major component of air (78%), is more soluble in wine than oxygen, though much less so than carbon dioxide. Nitrogen gas is used in stripping oxygen out of wines, especially just prior to bottling, a process in which it seems more effective than carbon dioxide. For blanketing wines in semi-filled storage containers carbon dioxide is preferable; being heavier than air, it tends to settle on the surface of the wine and is not as readily dissipated as nitrogen. (Commercial winemakers need to watch carbon dioxide absorption so it remains below the one atmosphere of pressure permitted in table wines—sparkling wines are subject to a much higher federal tax.) Nitrogen gas is sometimes used to flush wine bottles just before they are filled.

The presence of hydrogen sulfide gas is always a cause for concern. It seems to be mainly produced by yeasts during the reducing condition of fermentation; some yeasts such as Montrachet are more prone to form it than others. Sulfur sprays used in vineyards may contribute to its production. It has recently been suggested that yeasts starved for nitrogen-containing nutrients can exude an extracellular enzyme that breaks down sulfur-containing proteins and amino acids and forms hydrogen sulfide. The detection level of this gas in wine is about 1–2 parts per billion (ppb). It can usually be dissipated during the first few weeks after its production by racking and aerating. After several weeks, however, it tends to react with other components to form less volatile mercaptans, and in a few months these can be further oxidized to disulfides, which are almost impossible to remove by a stripping procedure. Winemakers should check frequently for hydrogen sulfide during fermentation and the first few rackings and take swift action to remove this noxious component if it appears.

Chemical Additives

Although various substances are added to wines at different stages, most of them, such as fining agents, are removed before bottling. Currently the only substances added that remain in the wine are preservatives or antioxidants. These include sulfur dioxide, ascorbic acid, sorbic acid, and fumaric acid.

The chemical structures of some preservatives are given below:

$$SO_2 \qquad K_2S_2O_5 \qquad CH_3—CH{=}CH—CH{=}CH—COOH$$

sulfur   potassium
dioxide   metabisulfite   sorbic acid

Enough sulfur dioxide to give 50–100 ppm is normally added to grapes at the time of crushing. A convenient source of sulfur dioxide is potassium metabisulfite, which reacts with grape acids to release sulfur dioxide. (If grapes are in poor condition, up to 200 ppm of SO_2 is sometimes used.) It is added as a gas in large wineries but in smaller operations a water solution of a sodium or potassium salt is more usual.

Sulfur dioxide has a variety of desirable effects. It protects wines against oxidation, deactivates enzymes that cause browning, retards the growth of bacteria, reduces oxidized smells, and improves the color of red wines. When dissolved in water or wine, it exists mostly as the bisulfite ion (HSO_3^-). Bisulfite reacts readily and essentially completely with acetaldehyde and to a lesser extent with other substances in wines. These products are called "fixed sulfur dioxide."

The portion of the total SO_2 not fixed in this way is called "free sulfur dioxide." The reaction of SO_2 with acetaldehyde and other compounds in wines is reversible, so as free SO_2 is lost through volatilization or oxidation, more free SO_2 is slowly released.

Sulfur dioxide has an unpleasant odor if present in excess, a fact that places an upper limit on its use. Gaseous SO_2 reacts with receptors in the nose causing pain and can also give an unpleasant, musty taste to wines. If total SO_2 is 60–150 ppm, free SO_2 will be about 15–50 ppm, and most people will not notice an adverse sensory effect. Small amounts of SO_2 actually enhance the smell of wines by removing the faded, oxidized odor of acetaldehyde. The total SO_2 concentration permitted in commercial U.S. wines is 350 ppm; only half this amount is permitted in dry red wines in Europe. It has been reported that a few people experience a severe allergic reaction to SO_2, and they should probably avoid wines containing it.

During fermentation some SO_2 is oxidized to sulphate, perhaps by enzymes. Active fermentation soon binds the free bisulfite because acetaldehyde is one of the intermediates in ethanol production. After fermentation, enough SO_2 should be added to combine completely with any remaining acetaldehyde. Winemakers should try to keep 5–40 ppm of free SO_2 in their wines at all times. A free SO_2 concentration at bottling of 20–30 ppm is usually considered adequate, although less can sometimes be used.

Sulfur dioxide kills many undesirable bacteria; it is effective against vinegar-producing bacteria, and as little as 100 ppm can inhibit malolactic bacteria. The antiseptic properties belong mostly to the dissolved gas, the amount of which is strongly pH dependent. At pH 2.8 it makes up 10% of the free SO_2; at pH 3.8 just 1%. This is one reason why high pH wines tend to spoil.

· Yeasts are somewhat sensitive to SO_2, especially in the presence of alcohol. High SO_2 levels can interfere with a good secondary fermentation in sparkling wine production. SO_2 is not as effective, however, in retarding wild yeasts as was once thought and alone will usually not prevent yeast growth in sweetened table wines.

The pigments in red wines loosely bind SO_2, a fact that does not affect the antiseptic action much but does make it difficult to analyze for free SO_2. Sulfur dioxide helps dissolve red grape pigments, and red wines with SO_2 usually retain their color better than untreated wines. But excess SO_2 can bleach wine pigments.

Ascorbic acid (vitamin C) and its isomer, erythorbic acid, have been added to wines to provide a type of antioxidant action. Ascorbic acid functions not as a true antioxidant but as a catalyst for the absorption of oxygen by other substances, especially sulfur dioxide.

When air contacts wine, certain phenolic compounds are converted to very strong oxidizing agents, which can oxidize ethanol to acetaldehyde and have other unwanted results. By itself, SO_2 does not react fast enough with oxygen to prevent some of these reactions, especially when much air contacts the wine quickly. Ascorbic acid increases sulfur dioxide's ability to absorb oxygen and is most effective when added just before wines are aerated by racking or filtering.

Ascorbic acid is tricky to use because without sufficient SO_2 present it can catalyze wine oxidation, so some wines treated with it undergo more oxidation than untreated wines. On the other hand, if too much SO_2 is present it may be oxidized to excessive amounts of sulfuric acid. If ascorbic acid is added at 60–180 ppm, the sulfur dioxide should be below 100 ppm. In most cases it is preferable to limit wine's contact with air rather than to use ascorbic acid.

Sorbic acid has been used to preserve sweet wines and prevent renewed yeast fermentation. It inhibits yeast growth but does not kill it so yeast levels must be very low for sorbic acid to be effective. The inhibitory effect is a yes-or-no proposition; if insufficient sorbic acid is used it is ineffective. Usually 150–200 ppm is required. The maximum allowed in the United States for commercial wines is 1000 ppm, but levels this high are never used. Sorbic acid is normally added in its potassium salt form since the free acid does not dissolve well in wine. The free acid is the most active form of sorbate and its concentration is greatest at low pH values. Alcohol has a synergistic action together with sorbic acid in preventing yeast growth. When used, sorbates are added just before bottling.

The taste threshold for sorbic acid in wines is about 135 ppm, though some people are sensitive to 50 ppm. This acid is odorless but it can result in an odor in wines because of either chemical or bacterial action. In wines stored for some time, it gives rise to an odor described as being like butter or oxidized fat.

When lactic bacteria metabolize sorbic acid, a strong odor resembling that of geraniums has been observed, so sorbic acid should not be used with red wines where a malolactic fermentation is possible. It is not effective against wine bacteria and should be used along with enough sulfur dioxide to prevent a malolactic fermentation. Unfortunately, sorbic acid reacts with and reduces free sulfur dioxide, sometimes permitting bacteria to grow. Although it is the only approved and effective chemical additive that retards yeast growth, its use in commercial winemaking is declining, and most amateur winemakers should probably avoid it.

Fumaric acid at about 500 ppm has been used by some wineries to prevent a malolactic fermentation. Malolactic control is helped by the presence of adequate SO_2 levels in the wine. In the author's experience, fumaric acid retards lactic acid growth but does not completely prevent it so where wild lactic bacteria are abundant, the addition of it may not solve the problem of gassiness and off odors in bottled red wines.

4
Winemaking Equipment and Materials

Winery Facilities

Few small winemakers have ideal facilities for making wine and compromises must often be made. Some features are more essential than others. During crushing and pressing, an easy-to-clean work area is desirable. A roof should shade workers from sun and rain. No one design is best for a winery work area. The main aim is to ease the amount of physical work needed by minimizing the distance wines need to be moved at each stage. There should be adequate lighting, hot and cold running water, and electrical outlets.

During fermenting, cellar treating, and aging the most important requirement is temperature control. A good location should have cool and relatively even temperatures year round, no direct sunlight, moderately high humidity if wines are to be aged in barrels, and no odors or fumes. Many small wineries are in basements, garages, or converted barns. In most of the United States a basement can be maintained at a fairly even temperature with a minimum of extra heating or cooling. Where basements are unavailable, special temperature control efforts are required.

Storage areas for equipment and supplies should protect against rusting, rotting, and other types of degradation. Garages and sheds can be used if the humidity is not excessive.

Necessary Equipment for the Home Winemaker

Suitable containers for fermenting small quantities of wine include 5-gallon glass carboys, 7-gallon plastic containers, and 1-gallon glass jugs. Glass carboys, available from shops selling home winemaking supplies and from companies that distribute bottled water, make ideal containers for fermenting white wines. Red wine fermentations, which require punching down the cap of skins several times per day, can be conveniently carried out in hard plastic containers. Such containers, sold by home winemaking suppliers, have tight lids that exclude air during the later stages of the fermentation and are easy to clean and store. One-gallon glass jugs are useful for very small fermentations and for handling small amounts of wine left over from larger containers. Gallon jugs can be obtained from bottle suppliers, or empty wine jugs can be used.

Glass containers are preferable to plastic for wine storage because plastic has a tendency to "breathe" and allow air in. Large containers have less wine surface-to-volume ratio and permit less oxidation than

Commercial plastic   Commercial glass

Homemade

Air locks

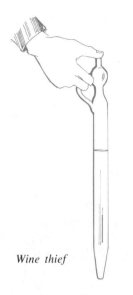

Wine thief

smaller ones. Quality wines can be stored in 5-gallon glass containers for several years, in 1-gallon glass containers for up to a year, and in small glass or plastic containers for a few months.

For the fermenting of white musts and pressed red musts, air locks are essential. These come in various forms, are generally made of plastic or glass, and fit into rubber stoppers (to seal carboys), into plastic fermenter tops, or into screw caps (to seal gallon jugs). Corks can be used to hold air locks but rubber stoppers are preferable since they do not have imperfections that allow air into the wine. Rubber stoppers (usually colored) sold for laboratory use should be avoided because they contain sulfur compounds as antioxidants, which can impart off odors to wines. Special rubber stoppers (usually white) sold by winemaking suppliers are what should be used for wine-making.

To siphon wines into clean containers after fermentation, rubber or plastic tubing is needed. Clear plastic tubing is preferable because it is easy to check for cleanliness and to clean. Clamps and valves are used to regulate the flow of wine during siphoning. Valves are usually more expensive than clamps but do not squeeze or damage tubing and permit more accurate control of the flow rate—valuable when one is finishing a racking or working with small containers. Special valves that permit wine flow only when pressed down on the neck or bottom of a bottle are available for bottling.

Dip legs are sold that attach to tubing and facilitate the siphoning of wines without stirring up the lees in the bottom of a container. Some of these have holes above the bottom and others are turned up in a "J." Another valuable piece of transfer equipment is a "wine thief," typically a glass or plastic tube that can be lowered into a container, stoppered with a thumb, and then removed to carry out a wine sample. A kitchen baster can be used to take samples but should be dedicated to this one use.

Kitchen equipment useful in winemaking includes funnels, sieves, measuring spoons (for the addition of dry chemicals in lieu of a scale), and measuring cups (for adding sugar). Pots can be used to catch juice from a press and to transfer small amounts of juice or crushed grapes between larger containers.

Bottles and carboys can usually be cleaned with a hot water rinse plus washing soda or suitable detergent. Stubborn deposits can be disloged with brushes that fit into bottles and carboys.

A variety of corkers are on the market including hand-held devices, bench models, and larger floor models. For anyone bottling several hundred bottles of wine per year, a good bench corker is a joy to use and a good investment. Cappers are useful in sparkling wine production and come in hand-held and bench models.

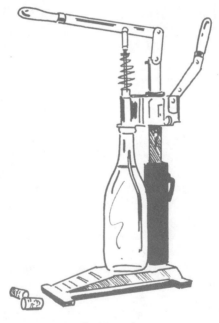

Bench corker

Several measuring tools are required to make good wines consistently. The most important are hydrometer, thermometer, acid test kit, and residual sugar test kit. All are available from most home winemaking suppliers. A hydrometer, which measures sugar in grape juice (or other starting ingredients) and fermenting wines, is invaluable for adjusting the initial sugar content to control wine alcohol content. A thermometer is needed to adjust fermentation temperatures and to monitor temperatures during malolactic fermentations and wine storage. Some means of measuring acidity is important in most grape wine production and all other wine production. A means of measuring residual sugar is needed to determine when wines are dry and stable and can be bottled and is also useful in checking the approximate sugar content of wines meant to be slightly sweet. The popular Dextrocheck® kit, available from many suppliers, can be used to estimate residual sugar up to 1% (or higher if the sample is diluted). For a simple yes/no check to see if a wine is dry, urine sugar test strips, such as Clinitest®, available in drug stores, can be used.

Other Useful Equipment for the Advanced Winemaker

Winemakers who deal with larger amounts of grapes or other fruits need at minimum a crusher and a press. Crushers for grapes should be of a type that breaks the skins without smashing the seeds. Crushers for apples and other firm fruits should be of a type that macerates the fruit pulp. Small hand-driven and larger motor-driven crushers are available in several styles. Hand-powered vertical basket presses are the mainstay of the amateur winemaker. They are slow and not efficient in extracting juice, but the juice they yield is of excellent quality, very low in suspended solids. Some models use hydraulic pressure, which increases efficiency.

A glass fermentation tube is a useful device for checking the rate of fermentations, especially slow ones (see Appendix B for details of its use).

Stemmer/crushers remove stems from grapes and greatly speed the processing of red grapes. They are available in various sizes, both hand-driven and motor-driven.

Filters, widely used to clarify and stabilize wines, come in many sizes and shapes, as do filter media, some common forms of which are sheets, cartridges, diatomaceous earth (D.E.), and membranes. Typical amateur systems that use paper sheet filters filter wines by gravity and require 10 minutes to an hour per gallon. Many of these systems do more harm than good because they permit excessive oxidation.

Cartridge filters have advantages in larger-scale home winemaking because the equipment is simple and easy to clean and yet has moderately fast flow rates. A typical system includes a pump, a chamber holding a 10-inch fiber-wound filter cartridge, and a pressure gauge to tell when a new cartridge is needed. Cartridges are relatively inexpensive. Filtering time for 5 gallons is about 2 minutes, and exposure to air is probably no more than that caused by a simple racking.

Diatomaceous earth filters use a special powder that builds up a filter coat on a support. A simple D.E. filter is more difficult to use than a cartridge filter, but for medium-large wine quantities it can be less expensive. Small commercial winemakers often favor this type. D.E. systems usually require some pretreatment of the filter media to remove tastes that would otherwise get into the wine. Some commercial winemakers start with water and add gradually increasing amounts of citric acid until the wash water reaches the pH of the wine to be filtered. With D.E. filtering systems one has to recycle the wine back into the original container until a D.E. coat builds up on the support and the effluent is clear. This step can lead to some oxidation and thus is a possible disadvantage.

Cellulose or plastic membrane filters are used for very fine filtrations, especially sterile ones. They generally require a skilled and

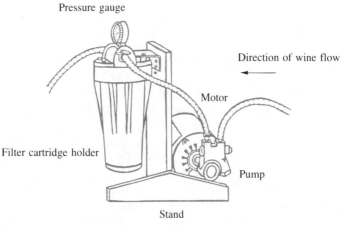

Cartridge filter setup

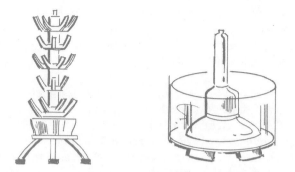

Bottle drying rack and bottle washer

careful user and are expensive. For most amateur and small commercial winemakers, cartridge and D.E. filters are most suitable.

Filters come in a variety of nominal pore sizes, where the stated size is more an average than an upper limit. For rough-filtering wines with sediment, a 5-micron filter can be used. For normal wines, one in the range of 0.6 to 1.0 micron is usually preferable. Mature yeast cells are generally removed by a filter with a pore size of 0.6 micron. Unfortunately, a sweet wine cycled through a 0.6-micron filter may referment for two reasons: (1) under the pressure of filtering (especially when a pump is used) buds on the sides of yeast cells can become detached and pass through the filter, and (2) yeast cells in the air can enter a wine after it is filtered but before it is bottled.

For a wine as clear as possible, filters in the range of 0.45–0.6 micron are used. True sterile filtration usually requires a membrane filter with an absolute pore size of 0.2 micron and elaborate technical skills. Nevertheless, by using a fine filter and carefully sterilizing bottles and other equipment a small winemaker can be fairly successful in preventing sweet wines from refermenting in the bottle. A good filtration also lessens the chances of unwanted bacterial action in bottled wine.

Oak barrels are widely used in aging premium wines and sometimes in fermentations. Wood bungs for sealing are being displaced by rubber or composition bungs, which give a better seal and are less likely to damage the barrel.

Two handy items among the equipment available for cleaning wine bottles are a rinser that attaches to a faucet and sends a powerful stream of water up into a bottle, and a drying rack resembling a Christmas tree that enables about 4 cases of bottles to be drained at once in a small space.

Equipment useful during bottling includes small automatic bottle fillers with reservoirs that dispense wine when bottles are pushed up against spring-loaded valves), devices that apply glue to labels, devices smoothing lead foil capsules, and heat-shrink capsules and heat guns.

A set of cork borers is useful for boring holes in corks or rubber stoppers so air locks or other devices can be fit into them. A microwave oven is very useful for heating small amounts of liquids rapidly with minimum fuss. It can be used to sterilize grape juice for building up a yeast culture, to boil wine samples before acid titration, to prepare Sparkolloid® fining mixtures, to sterilize sweet wines before bottling, and to soften and sterilize corks. Spare refrigerators are indispensable for storing unfermented grape juice and sweet reserves, and for conducting cool fermentations or cold stabilizations when outside temperatures are too high.

Tanks of compressed nitrogen and carbon dioxide gas have several uses. Nitrogen can drive off hydrogen sulfide and other bad odors from wines without oxidizing them and expel oxygen just before wines are bottled to minimize bottle sickness. Carbon dioxide is useful for filling containers before white wines are racked into them (to avoid oxidation) and for displacing air from partly full containers until the wine in them can be transferred. With the proper equipment it can also be used in making carbonated wines.

Various types of laboratory equipment are valuable to advanced winemakers. A refractometer can measure the dissolved solids (mostly sugar) in individual grapes and is very useful in preharvest vineyard surveys. Scales (kitchen and laboratory) accurately measure ingredients and chemical additives. A vinometer is a simple calibrated tube that can give a rough estimate of alcohol content in a dry wine; an ebulliometer is a more expensive and accurate device for measuring alcohol content. A Cash still (used with acid-measuring equipment) measures volatile acidity in wines. A pH meter has several uses: to measure the pH of musts and wines, to give a rough estimate of the relative amounts of tartaric and malic acids, and to give an accurate endpoint in acid titrations (especially useful for red wines).

Necessary Materials for the Home Winemaker

This section on materials gives a general idea of their uses; further details are best understood in the context of particular steps in winemaking, discussed in later chapters.

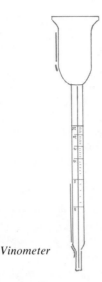

Vinometer

One essential supply for the winemaker is wine yeast (see Chapter 5 for details). Various clarifying and fining agents are used in winemaking to remove hazes. These include pectic enzymes, bentonite, and gelatin (see Chapter 6). Another useful material is an antifoam agent (usually a silicone oil that breaks bubbles) that can be added to fermentations in closed containers when foaming gets out of hand.

Modern winemakers realize the need for the judicious use of sulfur dioxide. One available form of it is Campden tablets, pills containing about 0.5 gram of sodium bisulfite. They need to be crushed and dissolved in warm water before being added to the wine, but are convenient when a balance for weighing is unavailable. One tablet per gallon of wine provides about 75 ppm of SO_2. Potassium metabisulfite is commonly used by small winemakers as a source of sulfur dioxide and is convenient as a 5% solution in water. Each milliliter of this solution added to a gallon of wine gives about 8 ppm of total sulfur dioxide.

Citric acid and tartaric acid can be added to increase the acidity of musts or wines when necessary. Calcium carbonate does the opposite. (Details of the use of these chemicals are given in Chapter 7.)

Certain chemicals are needed to clean and sterilize equipment. Washing soda (sodium carbonate), sold in grocery stores, is good for getting dried wine stains off glass and for cleaning used barrels. Dilute solutions of household bleach are helpful in disinfecting equipment but must be thoroughly removed. A sodium bisulfite rinse will help to neutralize any residual bleach. A water solution of about 1% sodium bisulfite plus 0.25% citric acid can be used for rinsing barrels and other equipment but is only a moderately effective sterilant.

For bottling wines, the common 750-ml bottle accepting a cork is generally preferable. Screw-cap bottles, gallon jugs, and 375-ml bottles are used in special cases (see Chapter 19). Sparkling wines require special heavy champagne bottles, and the preferred type has a lip that accepts a crown cap. Corks (see Chapter 19) and stoppers for sparkling wines (see Chapter 13) are necessary supplies, and labels and capsules to cover bottle tops are often desirable (see Chapter 19).

Other Useful Materials for the Advanced Winemaker

Malolactic bacterial cultures are used by many advanced amateur winemakers. (See Chapter 5.)

Other fining agents beside those mentioned above, are helpful in certain situations: egg whites and isinglass (fish protein)—preferred by some commercial winemakers for their gentle action on red wines—activated charcoal, casein, Sparkolloid®, and PVPP (see Chapter 6).

Various chemicals useful in advanced winemaking include potassium bicarbonate, for reducing acidity (see Chapter 7); copper sulfate, for reducing hydrogen sulfide or mercaptans; mineral oil, for removing mercaptans and disulfides (see Chapter 6); and potassium sorbate, for preventing refermentation of sweet wines (see Chapter 5). Granular oak gives an oak character to wines aged in glass or in old barrels, sulfur strips prevent spoilage in stored oak barrels, and various cleaning and preserving compounds are used in barrel maintainance (see Chapter 8).

Laboratory supplies needed by winemakers who routinely test wines include distilled water, sodium hydroxide, and phenolphthalein indicator. Other supplies are discussed in Appendix B.

5

Microorganisms
and Fermentations

A variety of microorganisms are involved in winemaking. Wine yeasts are fungi (saprophytes) that live on nonliving things. They convert sugar in grape juice into alcohol and carbon dioxide. Some fungi (e.g., molds that grow on grapes) are parasites and can attack living tissues. Bacteria, another type of microorganism, can be either helpful or harmful in winemaking.

Microbiologists keep revising their ideas of how various microbes are interrelated and from time to time recommend name changes. In this chapter some microbes important to winemaking are discussed; the names used are those currently recommended.

Molds

Molds do not grow in normal wines but can attack grapes and grow in wine cellars to affect the taste of wines. Downy mildew, powdery mildew, and various rots can reduce fruit quality and sometimes give off flavors. *Penicillium expansum* can attack grapes after fall rains have split them and may also live on corks and the insides of moist empty barrels. Even in small amounts it gives wines an unpleasant taste. This mold, which is often white, prefers cool temperatures.

Aspergillus niger also grows on grapes damaged by rains, especially in warmer climates. It is usually black and though it does not much harm the flavor of grapes, it may open the way to further infection by yeasts and bacteria. *Actinomyces* can grow in empty casks on or equipment and may give wines an earthy odor.

Botrytis cinerea, a mold known as "noble rot," has two possible effects. If the weather following its attack on the grapes is warm and dry, the mold sucks water out of the fruit, concentrates sugars, reduces acids, and leads to wines of superior quality. French Sauternes and German beerenauslese wines are produced in this way. But if weather following infection is cool and wet, *Botrytis* merely lowers grape quality.

Molds in vineyards can usually be held in check by a rigorous spraying program. Those on winery surfaces and equipment can be controlled with cleaners containing chlorine and general hygiene measures. Wet, empty wooden barrels can be kept effectively free of mold if sulfur wicks are burned in them every week or two.

Still Table Wine Yeasts

Yeasts important to winemaking include pure *Saccharomyces* cultures and also many wild strains. Normal wine yeasts are classified as *Saccharomyces cerevisiae* var. *ellipsoideus.*

Factors important in achieving clean yeast fermentations include: (1) yeast strain, (2) lack of competition from undesired yeasts, (3) innoculation volume, (4) yeast nutrients, (5) presence of air, (6) temperature, (7) additives in the wine, and (8) other wine components.

Wine yeasts are rare in vineyards where winemaking has been carried on for only a few years. In established winemaking regions the major yeasts are *S. cerevisiae* and *Kloeckera apiculata,* which live on fermentation residues spread on vineyards as fertilizer.

Although wild yeasts may be present at the start, as fermentation progresses the more alcohol-tolerant *S. cerevisiae* usually dominates. Enologists disagree about the possibility of improving wine flavors and fragrances by using mixed natural yeasts. Careful studies have shown that small fragrance and flavor differences can be produced, but these often decrease with time. Wild yeasts occasionally give undesirable odors. Various strains of *S. cerevisiae* isolated from different grape varieties make for minor flavor differences. Most flavor differences are due to the grapes used.

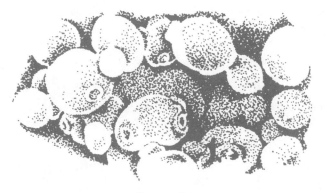

Yeast cells

American winemakers generally believe that natural yeasts are acceptable if they are well established in a vineyard but know that they grow more slowly than cultured yeasts and can cause problems. Therefore, rather than chancing a natural-yeast fermentation, most prefer pure cultures with known properties and results. Sulfur dioxide retards the growth of many wild yeasts, though it does so less than was once thought. In Europe, although natural mixed cultures are often employed, high sulfur dioxide in musts and the transfer of yeasts from vat to vat usually ensures that the fastest-growing yeast dominates.

Yeasts are available as dry or liquid cultures. Liquid cultures are built up by adding a growth medium (often pasteurized grape juice) and allowing the yeast cells to multiply. When a culture is vigorously fermenting, enough is added to grape must to equal 1–5% of the must volume. Various liquid yeast cultures are available from California laboratories in several container sizes (some suppliers are listed in Appendix D). Dry yeast, which has the advantage of being easy to store and use, is usually either added dry or after being soaked for a few minutes in tepid water but can also be built up. The usual addition is about 1 gram of dry yeast per gallon of grape must. Home winemakers can obtain a variety of natural yeast flora from European winemaking regions by using Vierka® brand dried yeasts. The mixture from each region varies from season to season and tends to be less vigorous than single yeast cultures but is suitable for experimentation. Yeasts of interest include a newly developed "cold fermentation" mixture.

Several producers supply pure dry yeast cultures. The following listing of Red Star® yeasts indicates the major types available:

MONTRACHET

S. cerevisiae, University of California at Davis #522. A general purpose, vigorous yeast for table wines having good flavor characteristics and high SO_2 tolerance.

PASTEUR CHAMPAGNE

S. bayanus, University of California at Davis #595. A moderately vigorous yeast used for sparkling wines and stuck fermentations.

CALIFORNIA CHAMPAGNE

S. bayanus, University of California at Davis #505. A moderately slow yeast that gives very compact lees for bottle-fermented sparkling wines.

FLOR SHERRY

S. fermentati, University of California at Davis #519. A yeast that gives high aldehyde production in both surface and submerged flor wines.

EPERNAY 2

S. cerevisiae, Geisenheim Research Institute selection. A slow-fermenting yeast, producing low foam, for use with table and sparkling wines.

At present only the Montrachet, Pasteur Champagne, and Flor Sherry yeasts are available in 5 gram packets.

Yeast cultures can be built up by using grape juice. Yeast grows faster at low sugar levels so juice diluted with an equal volume of water may be advisable at the start. Juice should be sterilized by being heated in a pressure cooker for 30 minutes or by being brought to a boil in a microwave oven. After it has cooled and a few percent of the yeast culture is added, the container should be plugged with absorbent cotton, stored at 70–80° F, and shaken frequently. The onset of fermentation can be observed and an estimate of its rate obtained by means of a fermentation tube.

Grape musts generally have adequate yeast growth factors. The addition of nitrogen to them is usually not necessary but may reduce the tendency of yeasts to form hydrogen sulfide (H_2S). In sparkling wine production, if the nitrogen content has been reduced too much during the first fermentation, added ammonium salts may speed bottle fermentation. For fruit wines it is usually necessary to add nitrogen and sometimes phosphorus. Urea supplies nitrogen, and diammonium phosphate supplies both nitrogen and phosphorus. Yeasts do not seem to have an absolute requirement for any vitamin, except possibly biotin.

In normal winemaking, yeasts ferment sugar without using oxygen (anaerobic fermentation) and largely convert it to alcohol and carbon dioxide. In the presence of air (aerobic fermentation), yeasts convert sugar to carbon dioxide and water and gain much more (12.8 times as much) energy for growth. Rapid multiplication of yeast cells at the start of fermentation is made possible by air dissolved in the grape must. After a must is innoculated with yeast, enough oxygen is usually available for the cells to multiply from 1 million/ml to 100–200 million/ml. Then, when dissolved oxygen has been used up, yeast growth slows and the conversion of sugar to alcohol and carbon dioxide becomes the main reaction.

Some winemakers transfer yeasts from one fermenter to another. The optimum time to transfer is when yeasts are most actively growing at about 15° Brix. Red wine fermentations are usually more rapid than white owing to higher temperatures and the fact that air is mixed into the must when the cap is punched down.

During fermentation yeasts are influenced by wine temperature and components, and they add heat and components to the wine. Strains of wine yeasts vary in their ability to adapt to low temperatures often preferred in white table wine production (as low as 35° F).

Yeasts generate heat during fermentation, and in large containers, cooling is often necessary. A rule of thumb says that without heat loss the temperature of must goes up 2.3° F for each degree the Brix goes down.

In addition to bisulfite added by the winemaker, during fermentation most wine yeasts produce 10–30 ppm of bisulfite by reducing sulfate in grape musts. Acetaldehyde formed during fermentation binds bisulfite in wine, leaving little or no free SO_2. At the end of fermentation the winemaker should add enough SO_2 to combine with residual acetaldehyde.

Aluminum, iron, and copper can inhibit fermentations, so wine should not come into contact with any equipment made of these metals. Just a few ppm of copper can also increase H_2S production during fermentation.

Yeast growth in grape musts affected with *Botrytis cinerea* may be inhibited by a substance produced by the mold but more likely the principal inhibitor is the higher initial sugar content.

Several factors influence yeast culture selection. Yeasts differ in the rates of fermentation they bring about and in their sensitivity to temperature and alcohol, their production of undesirable byproducts and foam, and settling properties. The main factor controlling fermentation rate, other than yeast strain, is temperature. Other factors are grape variety, °Brix, pH, and ammonia content of the must. At normal temperatures (65–75° F) most fermentations are over in about a week, but at temperatures of 40–55° F fermentations may last for 2–6 weeks or more.

Many strains of *S. cerevisiae* can ferment sugar to 16% or even 18% alcohol with a syruped fermentation where small portions of sugar are added after the initial fermentation slows down. A syruped fermentation is sometimes useful in getting the 15% alcohol necessary for making flor sherry (see Chapter 14). In Japan, sake (rice wine) is made by a process naturally simulating a syruped fermentation. The starch in the rice is slowly converted to sugar by *Aspergillus oryzae* mold, and *S. cerevisiae* yeast is able to produce 17% or more alcohol. The mixed culture of mold and yeast is called koji.

Most wine yeasts ferment glucose faster than fructose. Since fructose is sweeter, stopping a fermentation before a wine reaches dryness gives a higher fructose/glucose ratio and a sweeter taste than a dry wine sweetened to the same sugar content with cane sugar or grape concentrate.

Most yeasts are retarded by high sugar concentrations; they work fastest when sugar is about 1–2%. In the range of interest to winemakers, a sugar content above 25% definitively retards fermentation. German trockenbeerenauslese wines, made from musts containing 40–65% sugar, ferment very slowly and give final alcohol levels usually below 9%. One study suggests that 4.8% sugar retards fermentation as much as 1% alcohol.

Most U.S. winemakers prefer yeasts that yield the minimum of undesirable byproducts—H_2S, mercaptans, acetaldehyde, acetic acid, ethyl acetate, and higher alcohols—during fermentation. Some wild bacteria produce ethyl acetate in the early stages of the process. During the same time frame acetic acid is metabolized by yeasts and reduced. Hydrogen sulfide formation during fermentation has been attributed to yeast strain, elemental sulfur in the must, and lack of free amino nitrogen. Eliminating sulfur sprays close to harvest, properly clarifying juice, and adding diammonium phosphate to the must are methods of minimizing H_2S. Yeasts strains differ greatly in the amount of H_2S they produce; Montrachet appears to be a worse offender than other common strains.

Some winemakers add 2–6 grams of bentonite per gallon to fermenting juice, a step that reduces foam and proteins, though it tends to increase H_2S production. Cloudy musts and musts to which bentonite has been added ferment faster than settled, fined, or filtered musts. Rapid fermentation of white musts is generally undesirable, but overclarified ones may not ferment properly. Juice solids much below 0.5% often give slow fermentations and 0.5–2.5% solids are usually preferred.

Since many commercial yeast strains yield only minor differences in wine flavor, secondary characteristics (e.g., lack of foaming and good settling) often decide the choice of yeast.

Other Wine Yeasts

The yeasts used for sparkling wine production are often strains of *S. bayanus*. Most champagne yeast strains give a coarse, heavy sediment that settles rapidly after fermentation, minimizing yeast carryover during racking and greatly easing the clarification of sparkling wines after bottle fermentation.

A yeast strain used in Sauternes production is *S. bailii,* which ferments fructose faster than glucose (the reverse of what most yeasts do). The result is less sweet wines for the same percentage of sugar— an advantage, since Sauternes' high sugar content might otherwise be too cloying.

Spanish flor yeasts and similar types in the Jura region of France represent a special group. The former are currently classified as *S. capensis* or *S. bayanus*. The flor yeast strain most used in California is classified as *S. fermentati*. Flor differs from other wine yeasts in that it forms a film on the surface of wines containing 12–16% alcohol and oxidizes alcohol to acetaldehyde and other products that give flor sherries their character. On table wines such yeast growth is considered spoilage.

In experimental fermentations, *Hansenula* strains have shown some ability to increase wine bouquet and flavor.

Malolactic Bacteria and Malic Fermenting Yeasts

In many cool wine regions grapes are high in acidity, and the ratio of malic acid to other acids is also high. A malolactic (ML) fermentation, caused by certain bacteria, converts malic acid (a diacid) to lactic acid (a monoacid) plus carbon dioxide, thus halving the acidity due to malic acid. This special fermentation, desired in many red wines and a few whites, follows the fermentation of sugars. It generally increases wine complexity but reduces fruitiness. It can reduce total wine acidity by as much as one third. Many of the highest-quality red wines undergo ML fermentation. Winemakers in warm areas such as California often use it to increase spoilage resistance (because malic acid is removed) even when acidity is not excessive. (If acidity is lowered too much, the level can be raised again with tartaric or citric acid.) Winemakers should measure grape acidity and decide before the alcohol fermentation if they want to encourage or discourage a malolactic fermention, which often begins before the alcohol fermentation ends. This choice is best based on previous experience with the grapes in question.

ML fermentation sometimes improves wine flavor, probably by increasing complexity, but tasters cannot consistently detect flavor changes resulting from it. Reduction of acidity, lactic esters, acetoin, and diacetyl (which smells like heated butter) may account for much of the flavor change. Off odors are not unusual during ML fermentation but can normally be minimized by aeration at the end of the process. Since off odors and flavors are more noticeable in delicate white wines than in heavier reds, it is not surprising that German and some other white wine regions do not favor this step.

A winemaker who wishes to prevent ML fermentation should be very careful about sanitary measures in the winery and should not store wines in wooden cooperage that has previously held wines which have undergone it. New wine to be protected from ML fermentation should be removed from the yeast lees soon after alcohol fermentation and then either filtered or fined, treated to give 30 ppm free SO_2, and acidified to 3.3 pH or lower (if this can be done without making the taste too sour). Storage temperature should be kept below 60° F. The addition of 0.05% fumaric acid apparently has been successful in preventing ML fermentations in California. In commercial wineries, pasteurization or sterile filtration followed by sterile bottling is often used as a preventative.

To encourage ML fermentation the opposite treatments are used. This fermentation ideally begins before the alcohol fermentation ends. Wine should be left on the yeast lees longer than usual, the temperature should be kept above 65° F, and no SO_2 should be added. Because these conditions also favor other yeast and bacteria growth and spoilage, the wine must be watched very carefully for any signs of these. The general practice today is to innoculate with a pure strain of ML bacteria that has been isolated from successful mixed natural strains in the winemaking region. The main pure cultures of ML bacteria used in the United States are strains of *Leuconostoc oenos*. In California the *L. oenos* ML 34 strain and in the eastern United States *L. oenos* PSU 1 strain are widely used. Others commercially available include the Equilait culture from France and the 44–40 culture from California.

Many ML bacteria require special growth factors, not all of which are understood. They are known to thrive in the presence of autolyzing yeasts and seem to grow better in red wines containing some grape skin components. Apple juice diluted with 4 parts of water is a desirable ML culture medium. A grape-juice medium for larger starter cultures can be made by diluting 1:1 with water and adding 0.05% yeast extract. Lacking yeast extract, winemakers can add a small amount of compressed baker's yeast, which will supply nutrients for the growth of the bacteria.

The ML culture can be added to red wines at the time of pressing, ordinarily at about 5° Brix, and should be 1–5% of the volume of juice. A larger amount of starting culture speeds the desired ML fermentation and minimizes the possibility of spoilage by other microorganisms. A winemaker can encourage an ML fermentation in white wine by leaving the grape skins in the must for up to a day after crushing and before pressing. It is desirable to innoculate with an ML culture after alcohol fermentation has started because then the free SO_2 will be bound by acetaldehyde and other wine constituents. On the other hand, one should innoculate before the end of alcohol fermentation because the alcohol level will not be so high; after fermentation is completed it is difficult to speed an ML fermentation by adding a culture. Alternative methods are mixing several percent of a wine that has just completed a successful ML fermentation with the new partly fermented wine or racking the wine from the fermentation container into barrels that have held wines which have undergone clean ML fermentations.

Most researchers agree that measuring the titratable acid or pH is not a reliable method of keeping track of ML fermentation. Not only can bacteria produce lactic acid from carbohydrates in wine but there is also the possibility that potassium bitartrate may precipitate during the time that ML fermentation may be occurring. Paper chromatography (see Malic Acid in Appendix B for a source of directions) is probably the preferred method of following the course of ML fermentation, but enzymatic methods of malic acid analysis are also used by commercial wineries.

Despite all efforts, it is sometimes difficult to obtain the desired ML fermentation. In some cases it appears that necessary nutrients are missing from the wine, and in others it may be that bound SO_2 is inhibiting the bacteria.

At the end of ML fermentation winemakers generally start the finishing operations, which include aerating the wine to drive off bad smells, adding SO_2, reducing the temperature, and fining and filtering. Bacteria can be removed by a sterile filtration through a membrane filter with a pore size of 0.45 micron or less. If ML fermentation takes place after bottling, a haze, sediment, gas buildup, and acid reduction can result, along with an unpleasant sauerkraut smell.

Experiments with strains of *Schizosaccharomyces* (not one of the usual wine yeasts) which ferment malic acid to ethanol and other compounds, show some promise of simplifying winemaking. This type of fermentation could be an alternative to a malolactic fermentation. Some strains give wines with poor flavors but others are better. A mixed culture of a SO_2-resistant *Schizosaccharomyces* strain and *S. cerevisiae* has been suggested as a way to speed fermentation and minimize off flavors.

Genetic engineering experiments to produce yeast strains combining the desirable features of yeasts and bacteria may eventually result in yeasts that can simultaneously ferment sugar and malic acid.

Yeasts and Bacteria That Cause Spoilage

Candida vini and *C. valida* are film yeasts that grow on wines low in alcohol. They oxidize alcohol to carbon dioxide in the presence of air and if not checked can eventually reduce the alcohol level significantly. The author has seen a neglected red wine in a leaky container turned into something akin to soda pop by such film yeasts. Probably the best way to control them is to fill containers as full as possible, minimizing exposure to air. In larger containers, where space must be left for wine expansion during warm weather, sweeping air out of the headspace with nitrogen or carbon dioxide can help control film yeasts.

Dekkera are spore-forming yeasts that have been found in some California wineries, yielding a certain off odor and taste that has been described as "horsey."

Both wine yeasts and spoilage yeasts can cause serious problems in bottled wines including cloudiness, sediment, and dangerous pressure buildup in the bottle.

The most common bacteria responsible for spoiling wines are lactic bacteria, which grow best without air. These include the genera *Lactobacillus, Leuconostoc,* and *Pediococcus.* Several species may grow together during wine spoilage. Some lactic bacteria ferment glucose mostly to lactic acid while others produce carbon dioxide, ethanol, acetic acid, and glycerol as well and also convert fructose to mannitol. A wine spoiled by lactic bacteria usually has an unpleasant smell, haziness, and a sediment. Since many lactic bacteria can grow under conditions that favor a clean ML fermentation, however, the winemaker must take pains to favor those that are desirable and discourage undesirable types. If available malic acid is depleted with a clean ML fermentation and that is followed by racking, SO_2 treatment, fining, and filtering, a wine can be made relatively immune to other lactic bacteria growth.

A variety of acetic bacteria can, in the presence of air, convert alcohol to acetic acid and ethyl acetate. If wines of less than 15% alcohol are stored in containers that are not filled full or are not airtight, they tend to be converted to vinegar. Three species of acetic bacteria are found in wines: *Acetobacter aceti, A. pasteurianus,* and *A. peroxydans.* These strains are generally different from those used in commercial vinegar production. Typical wine spoilage organisms produce noticeable amounts of ethyl acetate and give a smell that can be described as like nail polish remover. Commercial wine vinegars with such a smell would be unacceptable. Acetic bacteria that grow in wines do not form a heavy film of the type found in cider; they may form a thin film but will often be distributed throughout the wine. High alcohol (>15%), low pH (<3.2), and 100–150 ppm of SO_2 effectively discourage acetic bacteria. Wines whose fermentation has stuck are especially susceptible to spoilage by these bacteria because without a fermentation no carbon dioxide is present to prevent air from getting at the wine in a partly filled fermenter. Pomace rapidly acetifies and fruit flies can carry acetic bacteria to exposed must or wine surfaces. The chemical reaction that results in acetic acid is:

$$C_2H_5OH + O_2 \longrightarrow CH_3COOH + H_2O$$

ethanol   oxygen   acetic acid   water

Much of the spoiled odor is caused by ethyl acetate, which is produced by acetic bacteria:

$$C_2H_5OH + CH_3COOH \longrightarrow CH_3COOC_2H_5 + H_2O$$

ethanol   acetic acid   ethyl acetate   water

A sound wine will not undergo acetification after bottling.

Lactobacillus trichodes, a spoilage bacterium that can grow in fortified wines of up to 21% alcohol, forms mannitol from fructose and also carbon dioxide, ethanol, acetic acid, and lactic acid. At pH levels below 3.6 its growth is retarded. Sulfur dioxide at 50 ppm retards such spoilage and at 100 ppm kills the bacteria. At one time in California and other wine areas lactobacillus bacteria caused a spoilage characterized by increased wine viscosity, due to polysaccharides formed by the bacteria as a protective covering. Wines and other liquids spoiled in this way became so thick that one could dip a finger into them and lift up a "rope"; hence the term "ropiness," used to describe the condition.

6

Cellar Operations—
Preventing and
Correcting Problems

What a winemaker does during cellar operations—racking, adding sulfur dioxide, fining, cold stabilizing, filtering, and so forth—has a large role in determining the style and character of wines and their sensory qualities: clarity, color, odor, and taste. Some problems in these areas can be at least partly solved later, but preventing them is usually easier and safer. This chapter can serve as a checklist of the more common problems winemakers encounter. All winemakers should keep careful records of their practices and results and identify recurring problems. They should then concentrate on learning about the causes of these problems and finding suitable preventative measures.

Preventing Clarity Problems

Hazes in wines can be caused by several factors: (1) suspended grape particles, (2) pectins, (3) proteins, (4) metal salts (especially calcium, iron, and copper), or (5) yeast or bacterial cells. The majority of suspended grape particles settle in time by gravity. Various fining agents can speed their removal. (Fining involves adding to a wine a material that settles by gravity, carrying down with it something the winemaker wishes to remove from the wine.)

Some grape and most fruit wines contain pectins (complex carbohydrates), which are colloidal and very difficult to get rid of by fining or filtering. Most grapes contain enzymes that break these down, eliminating this source of haziness, but other fruits by and large do not. Pectins can be removed from juice by the addition of a commercial pectic enzyme to crushed grapes or fruit. In many cases this step will not only improve wine clarity but will increase the juice yield. (See Appendix B for a test for pectins.)

EXAMPLE 6A
Use of pectic enzymes

A bushel of labrusca grapes (e.g., Delaware) or muscadine grapes (e.g., Scuppernong) is crushed and treated with 75 parts per million (ppm) of sulfur dioxide and with the amount of pectic enzyme recommended by its supplier. After the must has stood for 1–4 hours, it is pressed. (An attempt to settle and rack these juices is often unsuccessful.) The pectic enzyme generally increases both the ease of pressing these native grapes and the juice yield. The juice can usually be innoculated with a pure yeast culture immediately after pressing.

Another cause of wine haziness is protein, especially under hot storage conditions. This is usually less of a problem in red wines than in whites because the tannins in reds combine with and precipitate proteins. With white wines the most popular fining agent for removing proteins is bentonite. It is sometimes also used with reds; it removes particulates, some iron compounds, and essentially all proteins from wines.

EXAMPLE 6B
Fining with bentonite

Bentonite must be hydrated before use. One method of making a 5% slurry in water is to add 25 grams of bentonite to 500 ml of boiling water in a food blender, then blend this mixture for 30–60 seconds

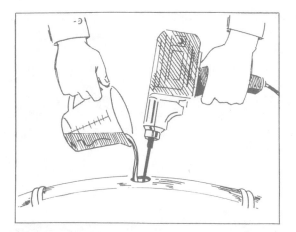

*Mixing in a fining agent with a stirring attachment on
an electric drill*

until a creamy suspension is formed. After this mixture has stood in a
bottle for a day, the bentonite is thoroughly hydrated and is an effective fining agent. It is usually used at a treat rate of 1–6 grams per
gallon of wine. (Some winemakers add it during fermentation but this
is not generally recommended because more bentonite is needed for
clarification and hydrogen sulfide formation can be increased.) Bentonite must be thoroughly mixed with the wine. Small containers can
be shaken by hand. For larger containers, such as 5-gallon glass carboys or barrels, one effective way of mixing is to use a stirrer on an
electric drill. After mixing, the bentonite will settle slowly and within
a week or so the wine can be racked away from it.

Adding small amounts of grape tannin has been advocated to help
clarify white wines. Excess tannins can give wines a harsh taste,
however, can darken a wine as they oxidize, and can form potent
oxidizing agents in the presence of air thus oxidizing other wine components, so tannins should be added in very small amounts on a trial
basis to begin with.

Iron, copper, and calcium salts can cause hazes. The first 2 types
are relatively rare since equipment containing these metals is now
seldom used for winemaking. Copper can get into wines if Bordeaux
mixture is used in vineyards too close to harvest time, so this practice
should be avoided. Adding about 1 gram of citric acid per gallon at or

shortly after fermentation prevents an iron haze even in the presence of iron. Calcium hazes can be a problem when calcium carbonate or additives containing calcium are used to reduce wine acidity. These hazes can be avoided if alternatives to calcium compounds are available.

Bacterial infections can also cause wine haziness. The main preventatives are hygiene measures and the use of adequate amounts of sulfur dioxide. Details are given below in the section on preventing microbial infections.

Preventing Color Problems

Too little color in red wines and two dark a shade in whites are the usual color problems. It is possible, but rare, for a red wine to be too dark.

Adequate red color can usually be attained if properly ripened grapes from a suitable growing area are used and adequate skin contact time is allowed for. If grapes have excessive skin pigments, skin contact time should be shortened to prevent too dark a wine color. To prevent excessive browning of white wines, sulfur dioxide should be used and air contact minimized during all stages of winemaking.

Preventing Loss of Fragrance

Methods of preventing fragrance loss include choosing properly ripened grapes from a suitable growing area, fermenting white wines at cool temperatures, avoiding malolactic fermentation, and excluding excess oxygen during the entire winemaking process. Certain yeast strains perhaps enhance fragrances but these are likely not to be natural grape fragrances.

Preventing Off Odors and Flavors

Major off odors found in wines include acetaldehyde, excess sulfur dioxide, hydrogen sulfide and related sulfur compounds, volatile acidity, and various odors caused by bacteria. The first two are most prevalent in white wines. They can be avoided by the use of the proper amount of sulfur dioxide and control of the amount of air reaching the wine.

The last three are mostly confined to reds. They can be prevented by the prompt racking of wines from their lees, by the use during fermentation of adequate sources of nitrogen and yeasts that are known to give minimum hydrogen sulfide, and by the proper use of sulfur dioxide.

Off flavors, especially in white wines, are often caused by permitting grape pulp and yeast cells to remain in contact with the wine too long. Many winemakers find that filtering white wines soon after fermentation and fining makes for wines with cleaner flavors.

Whether to filter wines is a controversial question to some smaller commercial winemakers, who believe that filtering removes desirable components. Others just don't want to be bothered by this extra processing step. Some European researchers believe that small particles of certain substances make positive contributions to a red wine's flavor. Manufacturers, suppliers, and many users of filter materials generally deny that there is flavor loss. Some believe that only the first portion of wine put through a filter loses flavor while the filter medium is absorbing some substances. Whether the pros of putting red wines through a fine-pore filter outweigh the cons will not be finally settled until definitive experiments are done. In the author's experience, when red wines are made in small batches they get enough air oxidation to convert some pigments into insoluble particles that have a bitter taste. Filtering such batches will probably improve flavor.

It is not good practice to filter wines before they have been fined because the large amount of suspended material in a newly fermented wine will quickly clog any filter medium.

Preventing Taste Problems

Major taste problems found in wines include too much or too little sugar, acid, or tannin, or a combination of these.

Winemakers aiming for a dry wine should either avoid grapes with more than 24° Brix or dilute the must with water to reduce the °Brix. A suitably vigorous yeast should be used, and the wine should be checked for dryness after several weeks before further cellar operations are performed. If a fermentation sticks at above 0.2% residual sugar, the wine should be racked with moderate aeration and stored at 60–70° F until fermentation finishes.

Winemakers desiring a semi-sweet wine should carefully monitor the fermentation and stop it when the desired sugar level is reached. Usual methods of stopping fermentation include chilling the wine below 40° F, racking it several times, adding up to 100 ppm of SO_2, and fining and filtering it. (Chapter 12 gives more details.)

Acid adjustments are dealt with in Chapter 7 because this topic is such a large one.

The major control for getting the desired amount of tannins in wine is adjusting skin contact time before and during fermentation. Tannin adjustments can be made later (the procedure is described later in this chapter).

Preventing Other Flavor Problems

Weak grape or fruit flavors can be avoided if fully ripe fruit grown in a suitably cool climate is used. One can avoid excessively strong grape flavors (such as those of some labrusca and muscadine varieties) by picking grapes before they are fully mature and by adding water. Baked flavors can be prevented by avoiding grapes from excessively hot growing regions.

Preventing Microbial Infections

Details on the variety of microorganisms that can cause spoilage in wines were presented in Chapter 5, so only a brief summary will be given here.

Molds do not grow in wines but do grow on equipment and inside empty or partly filled barrels. A winemaker can control them by using good hygiene practices and proper cleaning solutions (often containing chlorine) on equipment and storage containers, burning sulfur wicks in empty barrels, keeping wine-filled barrels topped up, and discarding barrels that molds have infected.

Undesirable wild yeasts can be controlled by the addition of a suitable amount of SO_2 to grape must and the innoculation of the must with a desirable yeast culture. Preventing air access to stored wine will avoid film yeasts. Yeast growth in sweet bottled wines can be eliminated or minimized by proper filtration (sterile if possible), pasteurization, or the addition of potassium sorbate at about 200 ppm.

A winemaker can minimize bacterial growth in wine by adjusting the °Brix of the must to get 12–13% alcohol; racking and clarifying the wine promptly after the alcohol fermentation; adjusting the pH as low as practical; keeping the free SO_2 at about 30 ppm; storing the wine in a cool place; encouraging a malolactic fermentation to remove malic acid; and either sterile filtration or pasteurization before bottling.

Quality Control Procedures

Winemakers should list the characteristics of available grapes or other starting materials, types of wine wanted, and possible problems in the production of those types. Quality control can be achieved by checking during each stage to make certain that the wines are progressing as planned. When potential problems are detected, corrective action should be taken promptly. Modern knowledge and techniques make it possible to correct many wine deficiencies. Home winemakers are more fortunate than commercial winemakers since they are not bound by sometimes dated regulations controlling wine treatments.

Restarting Stuck Fermentations

Reasons that a fermentation may fail to start or stop before completion include too weak a yeast, too much sulfur dioxide, too low a temperature, or too high a temperature. It is dangerous to permit a stuck fermentation to sit, especially at high temperatures, since without the reducing atmosphere and carbon dioxide present while fermentation is going on the must or wine is prey to bacterial infections.

Stuck fermentations must usually be dealt with by adding the stuck wine to a fresh yeast culture. Addition must be slow enough to prevent fermentation from stopping again. If too much sulfur dioxide has accidently been added to a must, it is usually advisable to rack the wine and aerate it to dissipate much of the SO_2 before attempting to restart fermentation.

EXAMPLE 6C
Restarting a stuck fermentation

Five gallons of white grape juice are innoculated with a special yeast strain. After 2 days there is little evidence of fermentation and a fermentation tube shows only a few small bubbles. A fresh culture of a more vigorous yeast strain is prepared and increased in volume to 1 pint with fresh juice. When fermentation is vigorous, 1 pint of the stuck must is added to the yeast culture. After this mixture is rapidly fermenting, a quart of the stuck must is mixed with it, and the winemaker continues to add stuck must to the fermenting mixture until the whole must is fermenting. With a vigorous yeast, this whole process may take just a day but it should not be rushed or fermentation may stick again and have to be started over.

Reducing Hazes

Hazes that appear after the initial fermentation and storage period can be reduced or eliminated by proper fining and filtering. Several fining agents are commonly used; with white wines the most popular is bentonite. Gelatin is often used to clarify reds (see the section on adjusting bitterness, below).

Fining improves clarity and sometimes flavor but filtering is often required for maximum clarity and stability. Even a small amount of haziness is very noticeable in white wines and filtering improves appearance. Many winemakers find that filtering whites as soon as possible after fermentation and fining have been completed gives a cleaner-flavored wine. The tannins in red wines combine with proteins and settle out naturally (or after a gelatin fining). Because of that and because red wines are dark, many winemakers do not bother with a filtration to improve clarity, but the step is often advisable to remove oxidized pigments and improve flavor.

It makes little sense to fine a wine and get it into good condition and then filter it and expose it to air. Most of the small filtering systems sold for home winemaking are so inefficient that they often do more harm than good. Typical amateur systems filter wines by gravity and require 10 minutes to an hour to filter just a gallon. A good filtering system can filter wine at a rate of at least 1 gallon per minute without excessive exposure to air. Faster rates are essential for commercial operations. (See Chapter 4 for more filtering details.)

Reducing Color and Off Odors

A variety of fining agents are used in special situations. Activated charcoal (at 1–5 grams per gallon) is used to decolorize or deodorize wines. Using too much carbon can remove desirable flavors, and in general carbon is used only to salvage wines that have major flaws. White wines are sometimes fined with casein to remove color, when, for example, a wine has been darkened by air exposure. One practice followed by commercial winemakers is to prepare a 5% casein solution and adjust the pH to 11 with ammonium hydroxide, then treat wine with less than 0.5 g of casein per gallon. Amateur winemakers sometimes use dried skim milk as a source of casein.

It is sometimes recommended that a casein fining be followed by a bentonite or Sparkolloid® fining (used at 0.5 to 2 grams per gallon), since casein by itself is not the most efficient clarifying agent.

EXAMPLE 6D
Fining with dry skim milk

For each gallon of white wine to be treated, 2 grams of dried skim milk are mixed with about 20 ml of water and then forcefully injected into the wine with a syringe. (Meat basters with needle points are appropriate.) The acids cause casein to coagulate almost immediately and if it is not adequately dispersed into the top layer of wine it will not have the desired effect as it settles.

After the coagulated casein has settled for 2 or 3 days, the wine is fined with Sparkolloid®, which is prepared by boiling 1 gram with 30 ml of water for 15 minutes (the water lost through evaporation must be replaced). The still hot mixture is added to the wine, and stirred in so that it is evenly distributed in the top layer. A week to 10 days later the wine is racked away from the combined sediment.

Since Sparkolloid® is positively charged, it combines with negatively charged particles and is effective in removing them from wines. Bentonite is negatively charged so Sparkolloid® can also be used as a "topping" agent for bentonite (and other finings) where it can assist in completing the precipitation and compacting the sediment.

Cold-mix Sparkolloid® is also prepared as a 3% suspension in water and used at the same treat rate. It is sometimes less effective than regular Sparkolloid® prepared by boiling but is easier to get ready and has some further applications. One application is its use along with pectinase enzymes prior to fermentation to clarify grape juice. Grape juice with lower suspended solids ferments more slowly and often gives cleaner-tasting wines.

Polyvinylpolypyrrolidone (PVPP) is a polymer powder that is wetted with a little wine before being mixed into a batch. Its main use is in decolorizing wines that have been darkened by oxidation. The normal treat rate is 0.5–2.5 grams per gallon. One commercial PVPP preparation is called Polyclar AT®.

Carbon, gelatin, and to a lesser extent casein and other fining agents can remove desirable flavors from wines. Casein and gelatin both have a softening effect because they combine with tannin, but when insufficient tannin is present, overfining can result. Sparkolloid® topping is recommended for certain "stuck" finings. Experience is the best guide to the proper use of fining agents.

Reducing Oxidation and Volatile Acidity

Oxidation is more of a problem in white wines than in reds since the tannins in reds act as antioxidants. On the other hand, red wines, because they are aged longer and often in barrels that are not airtight, are more subject to volatile acidity.

Acetaldehyde from minor oxidation can be eliminated if enough sulfur dioxide to combine with it is added. The amount of SO_2 should be selected after small-scale lab trials that allow the treated wine to rest for a day or two before it is smelled. The minimum amount needed to get rid of the oxidized smell should then be added to the main batch of wine.

In some cases, wine blending can be used to counterbalance the character of older, slightly oxidized wine. This technique is often used in the production of nonvintage French Champagnes.

More serious oxidation can be removed if the wine is added to a fermenting must of similar type or the pomace from a fermentation just completed. The yeast will reduce the acetaldehyde in the oxidized wine to alcohol. This procedure may fail, however, so the new wine should not be of such high value that its loss will be unacceptable.

Volatile acidity can also be reduced by the technique of adding wines to fermenting musts. The advantage here is that during the vigorous stages of fermentation the ethyl acetate associated with volatile acidity will largely be driven off. Volatile acidity is reduced proportionately with all acidity when wines are ion-exchanged—a useful fact to know when wines of slightly high volatile acidity are also high in total acidity. Ion exchange does not remove ethyl acetate, however, so if a significant amount of this sweet/sour smelling component is present, the treated wine may still be unacceptable.

Commercial winemakers usually handle a minor volatile acidity problem by blending wines high in such acidity with wines low in it. Home winemakers can do this, but should make trial blends and convince themselves that the result is not of detectably lower quality than the untainted blending stock. In case of doubt, it is better to discard wines high in volatile acidity rather than spoil more good wine.

Reducing Odorous Sulfur Compounds

One sulfur-containing compound occasionally present to excess in wine is sulfur dioxide. People's sensitivity to this substance varies; a few will find objectionable wines that contain as little as 50–100 ppm of it.

There are several ways to reduce sulfur dioxide. The usual procedure is to rack the wine with splashing to allow air to remove some of the compound. In general this method gets rid of only relatively small amounts. One can remove larger amounts by slowly adding small amounts of afflicted wine to a large batch of fermenting wine of the same type. The evolution of carbon dioxide will carry off some sulfur dioxide, and the acetaldehyde that is produced as an intermediate in the alcohol fermentation will combine with most of what remains.

An easier and more exact method is to oxidize the SO_2 with hydrogen peroxide. One reason for removing sulfur dioxide is to encourage a secondary fermentation in sparkling wines. Many yeasts find it difficult to grow in wines that contain 11–12% alcohol and 20 ppm or more free SO_2. Federal regulations permit adding up to 3 ppm of hydrogen peroxide to a wine for the purpose of encouraging a secondary fermentation (providing that none remains in the wine). This much hydrogen peroxide can oxidize 6 ppm of free sulfur dioxide.

EXAMPLE 6E
Lowering sulfur dioxide with hydrogen peroxide

To oxidize and remove 10 ppm of free SO_2 from a wine, the wine-maker adds 0.7 ml per gallon of ordinary 3% hydrogen peroxide (available from drug stores) and mixes it in thoroughly. Home wine-makers can add more hydrogen peroxide than this but should experi-ment with trial batches of wine before treating the bulk to be certain that they do not overdo the treatment. Measuring free sulfur dioxide before and after treatment is recommended.

Another sulfur-containing compound sometimes found in wines is hydrogen sulfide (H_2S), which smells like rotten eggs. Though it is usually formed by the yeast during fermentation, it may also be caused by the decay of dead yeast in wine lees.

A winemaker should smell wines every day during fermentation and every week or two thereafter to be certain that hydrogen sulfide is not developing. If it is detected, the usual treatment is to add some sulfur dioxide to prevent oxidation and rack the wine with plenty of aeration to drive off some H_2S. Another treatment that does not involve splashing the wine around is to treat it with a copper sulfate solution. The copper combines with the hydrogen sulfide to produce insoluble copper sulfide, which settles out. Federal regulations permit adding up to 0.5 ppm of copper to a wine provided that no more than 0.2 ppm remains. Copper should not be added during fermentation because at that time it can increase H_2S production.

EXAMPLE 6F
Removing hydrogen sulfide with copper sulfate

After fermentation has ended and the wine has been racked, a wine with a hydrogen sulfide smell is treated with a maximum of 0.5 ppm of copper, the amount contained in 0.75 ml of a 1% solution of copper sulfate pentahydrate added to a gallon of wine. After a day or two the wine should be smelled; if the hydrogen sulfide smell is gone, the wine can be treated normally. Normal rackings should remove essen-tially all the copper from it.

Another method of hydrogen sulfide removal is addition of the proprietary compound Sulfex®. This comes as a 10% slurry in water and is added to wine at the rate of 0.5–5 g/gal. It is insoluble, so it settles out without leaving a residue. It is also reported to remove traces of copper from wines and to stabilize pigments.

If H_2S remains in a wine longer than a few weeks, it can be converted into mercaptans, compounds that have a skunky odor and render a wine unacceptable. Because mercaptans are much less volatile than hydrogen sulfide, they cannot be removed by aeration. Treating the wine with copper, as for hydrogen sulfide, will remove them.

After several months, mercaptans can be converted into disulfides—just as smelly as mercaptans and even harder to eliminate. Neither aeration nor copper treatment will remove disulfides. Italian winemakers found a solution to this problem: they use olive oil to absorb the bad smells from wines, a method that appears to work because disulfides are more soluble in oil than in wine. But because olive oil can turn rancid and may itself contribute unwanted odors or flavors to a wine, a more practical oil is USP mineral oil available at drug stores.

EXAMPLE 6G
Removing disulfides with mineral oil

To a gallon of wine with a disulfide smell 1 oz of USP mineral oil is added. The bottle should be filled as full as possible to minimize air. The oil/wine mixture is shaken vigorously once a day for several days. One cannot tell when all the disulfide smell has been absorbed by simply opening the bottle and smelling because disulfides collect in the oil at the top of the wine. Instead, a sample should be taken from beneath the oil layer with a wine thief. When the wine smells clean, it is racked away from under the oil. If some oil is carried over, a second careful racking should leave these oil drops behind. The oil should not be reused for any purpose.

Adjusting Bitterness and Astringency

For red wines the most common fining agent is probably gelatin. Gelatin combines with tannins in wine and forms a precipitate, thus both clarifying the wine and lowering bitterness and astringency caused by the tannins. Home winemakers can purchase suitable un-flavored gelatin at a grocery store. It should be dispersed in lukewarm water and stirred occasionally for a half hour or so until no lumps can be seen in the mixture. Using boiling water or heating the mixture in any way is not advisable because this will tend to denature the gelatin and cause it to lose effectiveness.

The usual treat rate for gelatin is about 0.5 g/gal. Not only will gelatin remove certain particles from a red wine, it also removes an amount of tannin roughly equal to its own weight. In this way it can soften (make less bitter and astringent) the flavor of a young red wine and speed the time when the wine becomes drinkable. If a red wine does not contain much tannin (a rosé, for example, does not), it may be advisable to add some grape tannin before gelatin fining to avoid stripping too much tannin from the wine. (Tannins provide an antioxi-dant action and keep a wine from becoming too quickly aged.) Gelatin is sometimes also used with white wines, in which case it is necessary to add an equivalent weight of grape tannin (dissolved in warm water) first.

EXAMPLE 6H

Fining a red wine to reduce tannins

For each gallon of red wine to be treated, the winemaker softens 0.5 gram of unflavored gelatin in 10–20 ml of warm (not hot) water until all lumps have disappeared and a homogeneous solution results. This solution is mixed into about 4 oz of the wine to be treated and the diluted mixture is then thoroughly mixed with the bulk of the wine. After a week or 10 days, the wine should be racked away from the sediment.

Removing Trace Metals

Although trace amounts of iron and copper are rare in wines today, one may occasionally find a wine with more than 5 ppm of iron or 0.5 ppm of copper. One legal and satisfactory way to remove these is with the proprietary fining agent Cufex®.

EXAMPLE 61

Fining with Cufex® to reduce trace metals

Cufex® paste is diluted to make a 10% slurry in water and thoroughly mixed with the wine to be treated. The usual treat rate is 0.5–1.0 grams per gallon. This will remove 1–2 ppm of copper and iron (with copper being removed first). It is necessary to do laboratory tests to ensure that no excess Cufex® is used. After the Cufex® has settled, the wine is racked and filtered. Federal regulations do not permit a residue of more than 1 ppm of Cufex®.

Fighting Microbial Infections

The methods used to eliminate microbial infections depend on the type of microbe and the extent of infection. Early detection can minimize damage to the wine.

If normal wine yeasts begin to grow in what seemed to be a stable sweet wine, several courses of action are possible. The mildest treatment—in cases where just a little yeast sediment shows up and a few bubbles form—is to rack the wine, add about 50 ppm of sulfur dioxide, and keep a careful watch on it. If two rackings do not stop yeast growth, a tight filtration (through a 0.65-micron or smaller-pore filter, preferably a membrane filter) should be performed. Adding sorbates and pasteurization are methods of last resort since they often decrease wine quality.

Film yeast infections can be stopped by racking, adding 50 ppm of sulfur dioxide, and minimizing the stored wine's contact with air. Acetic bacteria can be eliminated in the same way. Lactic and other bacteria can usually be killed by keeping the wine cool (below 60° F if possible), adding 50–75 ppm of sulfur dioxide, and performing a tight filtration if needed.

Knowing When Corrections Are Hopeless

Off tastes and odors produced by yeasts, bacteria, and molds can seldom be removed from a wine. (An exception in some cases is volatile acidity.) Such spoiled or tainted wines should generally be discarded. But before doing so the winemaker should seek expert opinion to confirm that the off character is indeed due to micro-organisms. He or she should also try to pinpoint the source of the problem and take steps to ensure that it does not recur. With high standards of hygiene such problems should be rare. The major sources of microbial infection are probably wood cooperage and, in small commercial wineries, bottling lines.

7

Adjusting
Wine Acidity

All wines have a noticeable sour taste caused by natural fruit acids—tartaric, malic, and a few others. Because acidity greatly influences the taste of wines, winemakers need methods of measuring it, ways of relating these measurements to taste, and principles to guide them in adjusting acidity when necessary.

When an acid is dissolved in water or wine it partially separates (dissociates) into a hydrogen ion (H^+) and a characteristic anion (A^-). Many of the effects of acids depend on the concentration of the hydrogen ion. In wines these effects include influencing the sour taste, inhibiting the growth of unwanted bacteria and molds, and maintaining the color in red wines. Hydrogen ion and anion also instantly recombine into an undissociated acid until an equilibrium is reached. Chemists express this reversible relationship by the equation:

$$HA \rightleftharpoons H^+ + A^-$$

Strong acids (such as the sulfuric acid found in automobile storage batteries) dissociate almost completely, while weak ones (such as the acetic acid found in vinegar) dissociate only slightly (about 1%).

The effective acidity of a solution depends on both the concentration of acids present and their tendency to dissociate to hydrogen ions

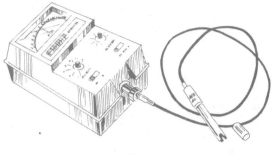

A pH meter

and is expressed by the pH scale. pH is defined as the negative logarithm of the hydrogen ion concentration in gram-atoms per liter. (A gram-atom of H^+ weighs 1 gram.) The pH of must or wine can be accurately measured with an electronic pH meter.

$$pH = - \log [H^+]$$

The H^+ concentration in most wines ranges from about 0.001 grams per liter (pH 3) to 0.0001 (pH 4). Water, considered to have no acidity, has a pH of 7.

Strong bases, such as sodium hydroxide, dissociate almost completely in water to give a metal cation and a hydroxyl anion.

$$NaOH \longrightarrow Na^+ + OH^-$$

sodium hydroxide	sodium ion (cation)	hydroxyl ion (anion)

Hydroxyl anions combine with hydrogen cations to produce water, thus neutralizing acidity. The other product of the reaction of an acid such as hydrochloric acid, HCl, and a base such as sodium hydroxide, NaOH, is a salt.

$$H^+ + Cl^- + Na^+ + OH^- \longrightarrow H_2O + NaCl$$

water   salt

The total acids in wine are usually measured with a titration. This process involves the measured addition of a strong base (usually sodium hydroxide) to a sample to combine with both dissociated (H^+)

and initially undissociated (HA) hydrogen ions. When free H^+ is neutralized, undissociated acids instantly dissociate to give more free H^+, and this continues until no more HA or H^+ is available. A titration to the neutral point (in practice about pH 8) measures all potential H^+, this giving the total acidity of both strong and weak acids. An example is the titration of tartaric acid. (In the following equation H_2T is shorthand for tartaric acid. Its full formula is given in the section on methods of increasing acidity, below.)

$$H_2T + 2\,NaOH \longrightarrow Na_2T + 2\,H_2O$$

Though half a dozen different acids are found in wines, it is traditional to express titratable acidity as though it were all due to tartaric acid, the major one. It is also the strongest and gives the lowest pH values for a given titratable acidity.

In winemaking, adding a strong base is not a good way to reduce

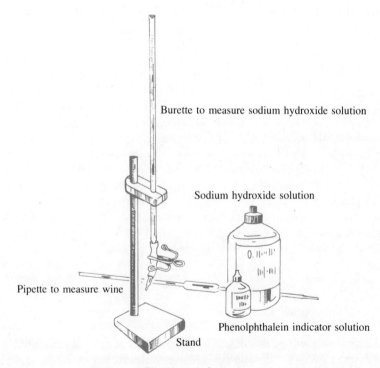

Burette to measure sodium hydroxide solution

Sodium hydroxide solution

Pipette to measure wine

Phenolphthalein indicator solution

Stand

Titration equipment

acidity. Adding a little sodium hydroxide merely causes more acid dissociation and the effective acidity is not much changed. By the time that enough sodium hydroxide is added to neutralize a substantial amount of acid, so much sodium tartrate or other salts will have formed that the wine will have an unpleasant salty taste and the pH will have risen dangerously high (permitting bacterial growth and giving brown colors). It is better to remove excess acids rather than to attempt to neutralize them.

For years there was uncertainty about the relative contribution of H^+ and HA to the sour taste in wines. Recent research has shown that about 10 times as much undissociated acid as hydrogen ion (H^+) is needed to give the same degree of sourness. But because HA exceeds H^+ in wines by about 100-fold, sourness is caused mainly by the undissociated acids. For this reason sour taste is more accurately predicted by titratable acid than by pH.

The results of tasting panels provide a quantitative measure of the relationship between acid taste, total acid, and pH. This relation, called the acidity index, is:

$$I_a = \text{Total acid (g/L)} - \text{pH}$$

In this relation total acid (titratable acid) is expressed in grams per liter. (Multiply percentage acidity by 10 to convert to g/L.) For most table wines, total acidity lies between 5 and 10 g/L and the pH between 3 and 4. The minus sign before the pH compensates for the fact that lower pH means higher H^+ concentration.

Calculations involving 350 commercial wines showed that for dry reds the acidity index averaged 2.50 and for whites 3.80. With whites the index was more sensitive to variety and to residual sugar. For example, Chardonnays averaged 3.20 and Rieslings 4.35. In general, sweeter wines had higher acidity index values. Although these preliminary numbers may be slightly modified by future research, they provide a guidepost for adjusting wine acidity.

Desirable Acidity in Wines

In dry table wines the desired level of total acidity is generally in the range of 0.6–0.75% and the pH generally in that of 3.2–3.6.

Wine type	Titratable acidity	pH
Dry white table	0.65–0.75%	3.2–3.6
Dry red table	0.60–0.70	3.2–3.6
Sweet white table	0.70–0.85	3.0–3.5
Semi-sweet red table	0.65–0.80	3.0–3.6
Sherries	0.50–0.60	3.4–3.9
Sparkling wines	generally the same as table wine of the same type, color, and sweetness	
Ports	generally the same as semi-sweet red table wines	
Fruit wines	generally the same as grape wine of corresponding style	
Flavored wines	within the ranges given above, depending on the use of the wine.	

An Outline of Acidity Adjustment Methods

Winemakers frequently need to adjust wine acidity when grapes are not of the proper ripeness. If must acidity is between 0.75 and 0.95%, in most cases little adjustment is needed to obtain the desired acidity in the wine. During fermentation, acidity often falls 0.1–0.2% as tartaric acid precipitates as potassium bitartrate. When must acidity is outside the desired range, the wine should be titrated after the first racking and adjustments planned if necessary.

Acid in wines can be adjusted by: (1) changing the number of hydrogen ions present, (2) changing the anions present, or (3) a combination of the two. In the following sections the methods listed here will be discussed in detail—first those to increase acidity, then those that decrease it, and finally the two that work both ways.

Method	Increasing acids	Decreasing acids
Via H^+ adjustment	Ion exchange	Ion exchange
Via A^- adjustment	Plastering	Ion exchange
Via H^+ and A^- adjustment	Add acids	Amelioration
	Blend wine	Cold stabilization
	Freeze wine	Blend wine
		Carbonate addition
		Acidex®
		Koldone®
		Malolactic fermentation

Methods of Increasing Acidity

"Plastering" is a process once widely used in the sherry district of Spain. Calcium sulfate (plaster of Paris) is added to the must, calcium tartrate precipitates, and sulfuric acid replaces tartaric acid.

$$CaSO_4 + H_2T \longrightarrow CaT + H_2SO_4$$

Plastering increases acidity and imparts a slight bitterness characteristic of Spanish sherries. Its disadvantages are that acidity and pH changes are not exact (calcium sulfate has low solubility and may not completely react), and too much bitterness may result.

Adding acids to raise wine acidity has the advantage that exact amounts of the acids desired can be added and pH adjustment is possible. The disadvantages of this procedure include the facts that (1) tartaric acid reduces cold stability, (2) malic acid increases the tendency for malolactic fermentation, and (3) citric acid can be converted by bacterial action to acetic acid.

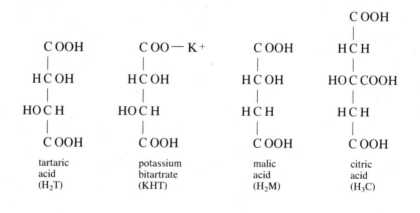

EXAMPLE 7A
Addition of acids

If the acid level must be raised, add citric acid at a rate of 0.85 gram per liter (g/L) (3.3 g or 0.12 oz per gallon) for each 0.1% (1 g/L) of acidity increase desired. Tartaric acid may be added instead, at a rate of 1 g/L (3.8 g or 0.14 oz per gallon) for each 0.1% (1 g/L) acidity increase, but some of the added acid may precipitate after addition.

Freezing wines and removing ice crystals increases alcohol, body, and flavors at the same time as acidity. No chemicals are needed and cold stabilization can be achieved at the same time. It is in some ways unfortunate that commercial winemakers are not permitted to freeze wines (in part because it is not a traditional practice), since freezing allows changes not achievable any other way.

EXAMPLE 7B
Freezing wine

A gallon of white wine made from a grape concentrate and containing 12% alcohol and 0.55% total acidity is placed in 4 wide-mouthed quart jars and stored in a refrigerator freezer compartment for 6 hours. The resulting slush is placed in a sieve and allowed to drain until about 80 ounces of wine is collected. The resulting wine has about 15% alcohol, 0.63% total acidity, and increased soluble solids and is suitable for use in a flor sherry fermentation.

The disadvantages of freezing include the need for a freezer or outdoor temperatures near 0° F and careful monitoring. Some wine is lost with the discarded ice crystals. Oxidation occurs during the draining process, making freezing unsuitable for most white wines but acceptable for dessert types such as sherries and reds.

Reducing Acidity by Amelioration

The acid index of wine made from underripe grapes or those grown in cold climates may lie above the desired range. In such cases, acids may be diluted, neutralized, or removed.

Amelioration, adding water to a wine to reduce its acidity and flavor, is a fairly widespread practice among commercial winemakers in the eastern United States. Its advantages include: (1) volume of the wine is increased by the amount of water added, (2) no chemicals or equipment are needed, and (3) strong flavors (such as those of Concord grapes or boysenberries) are reduced. Disadvantages include: (1) desirable flavors, color, and soluble solids are also reduced, and (2) acidity reduction is generally only about half as great as the water added (i.e., adding 20% water reduces the total acidity by only about 10%) because water increases the solubility of potassium bitartrate and less precipitates during and after fermentation.

Reducing Acidity by Chilling

Potassium in grapes combines with tartaric acid to form sparingly soluble potassium bitartrate. Alcohol decreases the solubility even further and some bitartrate precipitates from wines during fermentation. Since tartaric acid has two available H^+ s and the potassium ion replaces only one of these, the resulting potassium bitartrate is still acidic, and its precipitation during cold stabilization reduces wine acidity.

To ensure that the maximum amount of acid is removed in this manner, one must lower the temperature to near the freezing point of the wine (about 25° F) where the solubility of potassium bitartrate is much lower than at room temperature. Wines so treated are "cold stable," meaning that bitartrate crystals will not form later in the bottle if it is refrigerated or chilled. At the low temperatures used for cold stabilization, precipitation is slow and may require weeks. Introducing "seed" crystals of potassium bitartrate speeds the precipitation.

EXAMPLE 7C
Cold stabilization of wine

A gallon of wine is seeded with ¼ tsp cream of tartar crystals (available in grocery stores) and shaken to mix them in. It is then stored in a garage in winter when the temperature ranges from about 20° to 30° F. The wine is shaken every day to redistribute the crystals, thus speeding crystallization of more cream of tartar from the wine.

After 2 weeks the wine is racked from the crystals, with air contact minimized as much as possible to prevent wine oxidation.

Small wineries often chill their cellars by opening a door during the winter. If wines are in barrels, winemakers often roll the barrels back and forth to mix the contents. If the desired temperatures cannot be achieved naturally during the winter, refrigeration is one alternative. Another is treatment with the proprietary product Koldone® described below.

Chilling requires no chemicals but reduces acidity only in wines with appreciable tartaric acid and potassium. The reduction is generally small (often no more than 0.05%), requires controlled low temperatures, takes some time, and is difficult to monitor. The best way to monitor the process is probably with a conductivity meter, which measures the ions (including potassium bitartrate) in solution. Most winemakers routinely cold-stabilize white wines (which are usually chilled before serving) but can also use cold stabilization to trim the acidity of slightly too acid red wines.

Reducing Acidity by Chemical Additions

Although neutralizing by adding a strong base is less satisfactory than removing acid, certain weak bases perform a different function. They reduce acidity not by converting hydrogen ions into water but by precipitating acid anions and leaving a much weaker acid. Calcium carbonate ($CaCO_3$) removes tartrate anions and leaves carbonic acid, which then dissipates as carbon dioxide and water, leaving no residual acidity. Acid reduction by this method can be very accurate if sufficient tartaric acid is present.

$$H_2T + CaCO_3 \longrightarrow CaT \downarrow + H_2CO_3$$
$$H_2CO_3 \longrightarrow H_2O + CO_2$$

Another method for lowering acidity is the addition of potassium bicarbonate ($KHCO_3$), which gives the relatively insoluble potassium bitartrate (KHT), water, and carbon dioxide.

$$H_2T + KHCO_3 \longrightarrow KHT \downarrow + H_2O + CO_2$$

EXAMPLE 7D

Lowering acidity with potassium bicarbonate

If acidity is below 1.0% (10 g/L), it can be lowered by the addition of potassium bicarbonate. For each 0.1% (1 g/L) that the acidity must be reduced, add 0.9 g/L of potassium bicarbonate (3.4 g or 0.13 oz per gallon). The wine should then be chilled to complete precipitation of potassium bitartrate.

With wines having an initial acidity much above 1.0% there is a danger that adding enough potassium bicarbonate to lower the acidity to 0.7% will deplete too much tartaric acid and result in an unstable wine with a pH above 3.6. Trials should be done on small batches and pH measured before larger amounts of wine are treated.

Wines that cannot be satisfactorily deacidified with potassium bicarbonate are those with both high acid and high pH, usually because of high malic acid content. Malic acid can be precipitated when calcium carbonate is added to overly acid wines or musts:

$$H_2M + CaCO_3 \longrightarrow CaM \downarrow + H_2O + CO_2$$

Because calcium tartrate is less soluble than calcium malate and precipitates more easily, little malate is removed until nearly all the tartrate is precipitated. One method of getting around this situation is to treat only a small portion of wine with a large amount of calcium carbonate containing seed crystals of both calcium tartrate and calcium malate, thus forcing both acids to precipitate. This procedure is known as "double salt precipitation."

A proprietary mixture, Acidex®, with detailed directions for use is marketed to facilitate double salt precipitation. When properly used, Acidex® reduces both tartaric and malic acids in roughly equal proportions. In the final mixture excess calcium precipitates with remaining tartaric acid. This is probably the simplest and safest method of making significant acid reductions in grape wines. Under ideal circumstances acidity can be reduced by 0.7% or more.

EXAMPLE 7E

Lowering acidity with Acidex®

A winemaker desires to lower the acidity of 5 gallons of wine, standing at 1.2% with at least 0.4% tartaric acid, to 0.7%. The first step is to place 65 grams Acidex® in a mixing container and stir it with a pint of wine. Then 2.75 gallons of wine are slowly added while efficient stirring continues. After the treated wine has settled for several hours, it is racked from precipitated crystals and mixed with the untreated wine. The deacidified wine is allowed to stand for several months before bottling to permit dissolved carbon dioxide to dissipate.

It should be noted that Acidex® requires very thorough stirring and that malic acid reductions are almost always less than tartaric acid reductions. This is a fairly costly method and is generally not worthwhile for acid reductions of less than 0.3%.

Koldone® is another proprietary product (probably containing calcium carbonate and seed crystals of calcium tartrate) designed to cold-stabilize wines without cold temperatures. It may be of interest to winemakers in warm climates and to those wishing to avoid the effort needed to achieve the precise low temperatures required for traditional cold stabilization. If enough tartaric acid is present, Koldone® can also be used for moderate acidity reductions beyond what cold stabilization provides.

EXAMPLE 7F
Cold stabilization with Koldone®

Wine with a total acidity of 0.85% requires acid reduction and cold stabilization. A slurry is prepared by mixing 5 grams of Koldone® with 50 ml of water and letting the mixture stand for 24–48 hours. It is then added to a gallon of the wine and shaken vigorously 3–4 times the first hour, then once an hour for the next 6–8 hours. After 24 hours, the wine should be racked or filtered to remove precipitated crystals. The total acidity will be approximately 0.7% and the wine should be cold-stable.

A treat rate of 3.4 grams per gallon reduces total acidity by about 0.1%. Finer adjustments of acidity and cold stability are possible if laboratory trials, following the supplier's directions, are made.

Koldone® has some disadvantages: (1) it does not work well in wines high in potassium; (2) it requires a lot of attention for the first day of treatment; and (3) it introduces calcium, which decreases slowly. Commercial winemakers should monitor calcium to be certain that it is reasonably low before proceeding with other cellar treatments.

Reducing Acidity with a Bacterial Fermentation

Malolactic fermentation is commonly used to reduce excess acidity in red wines and some whites. Special bacteria convert malic acid, with two acid groups, to lactic acid, with only one, thereby cutting the acidity due to malic acid in half. This is the only method that can eliminate malic acid and, when properly conducted, it can stabilize wines against further bacterial action that could result in gassiness and off flavors after bottling.

As we saw in Chapter 5, malolactic fermentation is not always a good idea. Though in some cases it adds desirable complexity and smoothness, in others it decreases fruitiness and can cause unwanted flavor changes. Many commercial red wines produced in cool growing regions undergo malolactic fermentation. It is less used with white wines because it is often difficult to institute (grape skin contact seems to promote malolactic bacteria growth) and because the majority of white wines do not benefit from a reduction in fruity character. Chardonnay wines are one of the few exceptions here. (See Chapter 5 for more details.)

EXAMPLE 7G
Lowering acidity with a malolactic fermentation

A red grape crop has a total acidity of 1.1% and a pH of 3.1 just before harvest time, and a malolactic fermentation seems appropriate. Several weeks before harvest a pure freeze-dried or liquid malolactic culture is obtained and added to a 1:4 mixture of sterilized apple juice and water in an 8-oz bottle, stoppered with a plug of cotton, and held in a warm place. (Fresh cider from a mill will do but cider preserved with additives should be avoided.)

When the grapes are crushed, about 2% of the juice is drawn off and diluted with an equal amount of water. The apple juice culture is added and the juice placed in a container stoppered with an air lock.

After the bulk of the grapes are fermented and pressed, enough calcium carbonate (about 2.5 grams per gallon) is added to reduce the acidity to 1.0% and raise the pH to about 3.2, and the malolactic starter is mixed in. Depending on circumstances, the malolactic fermentation will take 1 week to 3 months. One can roughly follow its progress by observing bubbles in the air locks after the wine is free of sugar and by titrating for total acid content. Malolactic completion is best confirmed by means of paper chromatography. The final wine acidity depends on the initial malic acid content of the grapes but should be in the range of 0.7–0.75%. After cold stabilization the acidity should be close to optimum but can be adjusted as needed.

Ion Exchange

Ion exchange can be used either to raise or to reduce the acid level. It differs from other acid adjustment methods also in that the material added—a special resin usually in the form of beads—is insoluble in the wine. Absorbed on the resin are ions, such as hydroxyl, which can be exchanged with ions in the wine, such as tartrate. Ion exchange is best done by passing wine through a column of resin. Commercial winemakers have used this method to replace potassium ions with sodium ions. Since sodium bitartrate is much more soluble than potassium bitartrate, this process cold-stabilizes wines without the usual chilling process. But because high sodium levels can change wine flavor and adversely affect people on low-sodium diets, it is not used much today.

Although raising wine acidity by exchanging H^+ ions for potassium (K^+) and other cations is possible, it is seldom done, largely because it removes much of the color of red wines. This procedure can be used to analyze wines for total cations because the increase in moles of H^+ (determined by titration) will equal the moles of all other cations that were exchanged. (A mole of one compound contains the same number of molecules as a mole of any other because the weight of a mole in grams is defined as equal to the molecular weight.) It is more usual for home winemakers and small commercial operations to employ ion exchange for reducing acidity rather than for raising it.

The advantages of ion exchange are: (1) wines of very high acidity can be reduced to the desired acid level, (2) all acids in the wine are reduced equally, and (3) the ion-exchange resin can be reused repeatedly, so the only cost for chemicals is for sodium hydroxide, which is relatively cheap.

The method's disadvantages are: (1) initial costs, for the resins, are higher than those for other methods and there are fewer suppliers; (2) the technique is somewhat complicated; and (3) some decrease in wine flavor is possible. Ion exchange is probably the method of last resort to alter wine acidity for the amateur and small commercial winemaker, but it is a technique that can work when others fail.

EXAMPLE 7H
Lowering acidity with ion exchange

A 4-gallon batch of white wine has a total acidity of 1.4%, which needs to be reduced to 0.7% to be palatable. Other methods either will not reduce acidity sufficiently or will result in a wine with high pH that is biologically unstable.

A glass column, approximately $1\frac{1}{2} \times 36$ inches with a stopcock on the bottom (sold by laboratory suppliers as chromatography columns), is plugged with a little glass wool at the bottom and filled with a slurry of a pound of strong anion exchange resin, such as Dowex A-7D, mixed with water. The resin slurry should be added in portions, with the stopcock opened just long enough to drain excess water after each addition without allowing the resin to go dry (air bubbles among the resin beads will give a very uneven flow of liquid through the column). When the column is filled, the resin is rinsed with water until the effluent from the column is odorless and tasteless. The column is then rinsed with a 1% solution of sodium hydroxide to convert it entirely to the hydroxyl form (OH^-). This conversion is complete when the effluent from the column has a pH of about 13. (Approximately 2 gallons of 1% sodium hydroxide will be needed.) The resin should then be rinsed with enough water to remove excess sodium hydroxide (the pH of the water should fall to under 8).

The wine to be deacidified is passed through the column until the pH of the effluent wine drops below 6. About 2 gallons can be totally deacidified. When they are blended with 2 gallons of untreated wine, total acidity should be about 0.70%.

After the resin is rinsed with water, it can be reconverted to the (OH^-) form, rinsed again, and stored wet in a closed jar, to be reused indefinitely.

Adjusting Acidity by Blending

Blending wines to adjust acidity is often the method of choice. Color and wine flavors can be adjusted at the same time. There is no cost for chemicals and no worry about chemicals or bacteria changing the odor or flavor.

A winemaker will have to plan ahead for suitable wines to blend. In some cases two wines, which are cold, hot, or biologically stable by themselves, will become unstable in a blend. Winemakers should make trial blends (a gallon or less) and observe them for at least a month before attempting full-scale blends. Successful blending is a craft and art that requires experience for success. Without taking care a winemaker could balance acidity but unbalance other wine features. (See Chapter 9 for more details.)

8
Aging Wines

A variety of changes take place during aging that affect wine quality. Slow oxidation and reduction reactions cause the principal ones; wood extractives can also have a significant effect on the odor and flavor of wines aged in oak.

The rate of maturation varies with wine type, extent of aeration, storage container, and temperature. The higher the content of tannin, sulfur dioxide, and other reducing substances in a wine and the lower the temperature, the slower aging will be. If the aging process (primarily due to oxidation) is too rapid, the wine will not develop its finest qualities; if it is too slow, costs increase, as does the risk of undesirable changes or contamination.

Most white wines are aged in glass, stainless steel, or other inert containers. A short aging period in oak (2–12 months) is used for some white types, such as Chardonnay. White wines bottled young (one year old) have a lighter color, more fermentation aroma, and less bouquet and barrel flavor, and taste fresher and fruiter. Quality declines with longer time in cask, though if some wines are bottled too soon they still contain yeasty odors. The lightest wines reach their maximum quality after 6 months to 2 years in the bottle; heavier and sweeter types may continue to improve in the bottle for a number of years. Many drinkers of common wines prefer the somewhat "raw" flavor of new reds to the mellow flavor of aged ones. Light red wines low in alcohol are probably best when drunk young. A winemaker should consider consumers when making decisions about aging.

Many wineries hold red wines in large containers for up to a year and then store the finest ones in oak for 1–3 years before bottling. The best red wine quality generally results from the use of small barrels and relatively low storage temperature (50–60° F). Many red wines improve in flavor and bouquet during 1–4 years of wood aging, but after a wine has reached its optimum quality it declines. On the other hand, if a red wine is bottled before it has aged properly, it will remain harsh in flavor and age very slowly in the bottle. Bottle aging supplements barrel aging but in general cannot take its place.

Heavier red wines that can benefit by aging in wood should be low in volatile acidity, have a moderate alcohol content (about 12%), and a total acidity of at least 0.6%, be free of off odors, and have a well-balanced and distinct flavor. Aging cannot improve a basically poor wine.

Controlling Oxidation

It has been recognized at least since Pasteur's time that normal aging of many red wines requires limited oxidation. Preventing excessive oxidation is essential. When the process is controlled, the quality of wines aged in different types of containers can be equally high. The primary method of control is to fill containers as full as possible, leaving only enough room for expansion due to warming.

Sulfur dioxide slows oxidation and aging. As each 4 parts of sulfur dioxide are oxidized, they remove 1 part of oxygen from the wine. Some winemakers burn sulfur wicks in partly filled barrels, but since this greatly increases the sulfur dioxide content of the wine, retarding aging, and since dry barrel sections can still leak air, this method has nothing to recommend it.

Oxygen makes up 21% of air and enters wines exposed to it. Unless special precautions are taken, oxygen is absorbed during pumping, racking, fining, and filtering. More enters when wine is aged in a not completely filled container or a barrel not totally tight, and when a container is opened for the contents to be sampled or topped up.

It takes 6–7 ml of oxygen to saturate a liter of wine. Cool wines, such as those undergoing cold stabilization, can contain this much dissolved oxygen. Wines absorb and consume oxygen very rapidly, at least 50 ml per liter from the air in a partly full container within 24 hours.

Research suggests that oxygen can chemically combine with wines in at least two ways. If only a small amount of oxygen is available to a wine, it is absorbed by metallic ions and oxidizes them. These ions then oxidize other wine components, perhaps acids. Later in the cycle, tannins, pigments, and sulfur dioxide are oxidized. This whole process can take years. Oxidized tannins and pigments are eventually converted to insoluble compounds and precipitate. Such precipitates are found in bottles of old red wines, but those of moderate age (up to 5 years or so) seldom contain them unless they also contain unstable pigments.

When oxygen is introduced into a wine in excess or too rapidly, metallic ions cannot act as its primary carriers and the oxygen appears to combine directly with ethanol and higher alcohols to form aldehydes. The first result is a flat flavor in the wine. Such direct oxidation frequently occurs during bottling, with the result known as bottle sickness. It usually goes away in a few weeks, provided that the oxidation has not been too severe. Excess oxidation gives white wines a sherry character, unwanted in table wines.

Some higher alcohols in wine may be oxidized to acids, which can form esters with ethanol and other alcohols and contribute to the flavor and bouquet of aged wines. In small quantities some esters may lend a certain fruitiness to wines. Ethyl acetate, the principal ester formed in wines exposed to air, is undesirable in excess because it gives wines a chemical smell like lacquer thinner.

Wines from grapes with pronounced aromas and flavors tend to lose both as oxidation occurs. Oxidation causes the pigments in both red and white wines to turn brown and eventually precipitate, darkening white wines and lightening reds. Red wines can generally stand more oxidation and aging than whites because their tannins and pigments act as oxygen absorbers or antioxidants. If acetaldehyde is formed, tannins and pigments tend to combine with it to prevent a flat or maderized (referring to oxidized character, as is typical of wines from the Madeira Islands) aroma or flavor. Tannins have a somewhat bitter flavor that oxidation reduces, tending to smooth or mellow the taste. The dark color of red wines hides minor browning and color changes that accompany oxidation.

It is because aging in oak barrels is generally associated with oxidation that few white wines are left for long in oak. Exceptions are those of full body that can withstand some flavor loss and some darkening. Many white Burgundy, Rhone, and Graves wines are oak-aged in France.

The exact amount of oxygen required for proper aging is not

known. During fermentation and early rackings, wines contain carbon dioxide (which prevents oxygen from entering) and sulfur dioxide (which can neutralize absorbed oxygen). Some research suggests that wines from hybrids do not respond to oxygen in the same way that wines from viniferas do. It is estimated that red wines require at least 100 ml of oxygen per liter, absorbed slowly over a period of years, and that white wines need perhaps one-tenth as much. Lighter, more delicate red wines develop better and have longer shelf life if handled more like whites. A difficulty in measuring the exact amount of oxidation required is that little is known about the effects of reduction in the later stages of barrel aging and following bottling. Some research has shown that phenolic components of red wines oxidize more easily the second time (after an intermediate reduction) than they do the first. Until more is known about wine aging, the amount of aeration wines should receive is best judged by experience.

In barrels, wines have been reported to absorb about 30–40 ml of oxygen per liter per year and 5–6 ml during each racking. Some filter aids also increase oxidation, as do certain fining agents such as gelatin and charcoal. Heating, as during pasteurization, also increases it. Light and heat speed oxidation. Though rapid oxidation is undesirable for table wines, heat is widely used to oxidize dessert wines such as sherries.

In theory it should be possible to simulate oxidation during barrel aging by introducing controlled amounts of oxygen to wine in an inert container, but little research of this type has been done. The main method of controlling oxidation at present is the use of containers of different sizes (with different surface-to-volume ratios) and different porosities.

Inert Aging Containers

Since glass and stainless steel containers are airtight, usually the only precaution one need take against excessive oxidation is to see that they are filled as full as possible and that closures are also airtight. Wines stored in inert containers of 5 gallons or more are less likely to oxidize than wines stored in smaller containers. It is difficult to keep wines in good condition for more than a few months in containers holding less than one gallon.

If a wine being put into storage still contains traces of fermentable sugar, closing the container with an air lock is safer than using a solid closure. Closures on all storage containers should be checked frequently. Rubber stoppers easily loosen if the slightest pressure builds up in a container, and screw caps with plastic cone-shaped liners are also prone to loosen with time.

If a container cannot be completely filled, an inert gas (nitrogen or carbon dioxide) blanket can be used to protect against oxidation and acetic bacteria growth. Carbon dioxide is often preferred because it is heavier than air and has less tendency to escape if a small leak occurs in the container seal. It also seems to enhance the fresh character of young wines, the type most likely to be stored in inert containers. It is also desirable to fill empty containers with carbon dioxide before racking or pumping wine into them. As long as the pressure of dissolved carbon dioxide in a wine does not exceed one atmosphere (a term referring to the pressure exerted by air at sea level, about 15 pounds per square inch), no extra tax need be paid on commercial wines.

Inert containers can often be adequately cleaned with hot water, especially if one sound wine is being immediately replaced by another as during racking. More thorough cleaning calls for detergents and a brush for resistant deposits. A laboratory detergent such as Alconox® is satisfactory and rinses clean. Inert containers contaminated with bacteria or mold should be sterilized with bleach. Other details about cleaning containers are given at the end of the section on necessary materials in Chapter 4.

Aging in Wood

Good winemaking practice calls for cleaning up a wine before putting it into cooperage for aging, to prevent excessive growth of microorganisms or precipitation of materials inside a barrel. Also, the sooner a wine is stabilized and clarified, the better its final quality usually is. Wines destined for oak aging should be sound, without bacterial infections or other serious problems: they can be moved into barrels or casks as soon as they have settled fairly clear.

Fining or filtering red wines before oak aging may be unnecessary since oak tends to speed clarification. Moreover, wines fined too heavily may not age as well as they should. When wine is aged at a moderately low temperature, tartrate precipitation is speeded up, the wine clears faster, and evaporation losses are reduced. But storage at too low a temperature (below 50° F) slows aging.

During the aging of red table wines the volatile acidity should be periodically checked. High volatile acidity is probably the major defect of red wines aged in barrels. If it rises too high, 50–75 ppm of sulfur dioxide should be added or the wine should be sterile filtered or pasteurized. In no case should acetic bacteria be allowed to contaminate and spoil a barrel.

Wines in different barrels often age at different rates so it is a common practice to blend them together prior to bottling.

Wood Cooperage

For a long time wooden containers were the most common wine storage vessels. Under proper circumstances they are still preferred for aging certain wines. Small wooden containers are, however, the most troublesome and costly to use.

Disadvantages of wood cooperage include: (1) high initial cost, (2) high maintenance cost, (3) increased work to inspect and top barrels, (4) possibility of leakage, (5) possibility of bacterial contamination, and (6) loss of wine through evaporation. Stainless steel, glass, fiberglass, and other inert containers rarely have these disadvantages. On the other hand, many of the world's finest red table wines, some white wines, and many dessert and appetizer wines owe a good share of their quality to barrel aging.

Preferred sizes of wooden containers for aging run from 50 to about 500 gallons. Smaller barrels, sometimes used by home winemakers, have a higher surface-to-volume ratio that speeds aging and pickup of wood extractives. They are, however, more expensive to use than larger barrels.

Many different woods have been used for wine storage. Primary requirements are strength, resilience, workability, freedom from defects, and the absence of undesirable color or flavor extractives. The wood chosen should be widely available so that standardized cellar practices can be used for years. Given these requirements, the consensus is that white oak is the only practical wood for small cooperage (less than 500 gallons).

In barrel-making, an oak tree is cut into sections of the length desired for staves, and the sections are split lengthwise through the center into quarters. With only the inner heartwood being used, bar-

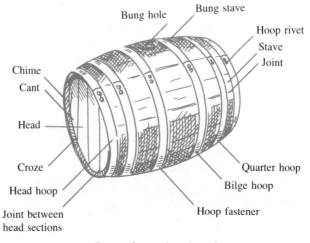

Parts of a modern barrel

rel staves are cut from the faces of each quartered section so that the rays in the wood, which extend outward from the center, are parallel to the width of the stave. Cutting them this way reduces wine diffusion through the staves and minimizes changes of stave width due to moisture.

What is the best white oak for wine barrels has been extensively debated. Oak qualities vary from tree to tree and within a tree, and also with species, growth conditions, and wood seasoning. A typical American 50-gallon barrel contains 31 pieces of wood, a random mixture from different trees.

The amount of white oak used for cooperage in the United States has greatly declined in the past 50 years, and barrel makers say that no shortage is imminent. The situation in Europe, where barrel makers must compete with furniture makers for limited amounts of wood, is different, and the prices of European barrels have risen sharply in recent years.

In the United States most of the oak barrels made are designed for aging bourbon whisky. Thus American wine barrels are usually bourbon-style barrels, which have not been charred inside.

The principal American oak species used in barrels is *Quercus alba*, which grows over much of eastern North America from southern Canada nearly to the Gulf of Mexico, and as far west as Texas. The European oak species (ranging over most of Europe) used for barrels are *Q. robur* and, to a lesser extent, *Q. sessilis*.

Preparation and Maintenance of Barrels

Letting a new barrel stand full of water for a day will tighten it up. A few hundred ppm (parts per million) of sulfur dioxide solution acidified with 0.25% citric acid helps to sterilize the barrel and prevent bacteria or mold from growing during the tightening process, but this soaking should not be too long or oak extractives will be lost.

Wine literature is filled with advice on barrel preparation. Some writers recommend that new barrels be treated with a hot 1% (or even higher) solution of sodium carbonate and allowed to soak for a few days, then rinsed with a 1% citric acid solution, then rinsed again with water. The usual rationale is that barrels need such a treatment to remove "excess" extractives, which would otherwise "taint" wine. It has also been suggested that low-quality wines be stored in oak barrels for a year or more before high-quality wines are placed in them. But such treatments serve only to remove valuable extractives from a barrel, lessening its value, and should be avoided by cost-conscious winemakers.

The outside of new barrels should be protected by a coat of linseed oil. Rotenone added to the linseed oil discourages cask borers. Alternatively, a strong hot solution of alum can be applied to the outside and allowed to dry before the linseed oil application. Hoop rusting can be prevented by protective coatings. The proprietary preparation V.E.X.® can remove mold, grime, varnish, and linseed oil from the outsides of older barrels.

After barrels have been used to store wine, ideally the winemaker removes the wine, checks (by smell or other test) to make sure that no bacteria or mold infection has started, and immediately fills the barrel with another wine. Tartrates and other sediments should be removed at least every 3–4 years. Cold water rinsing removes the least extractives and may be satisfactory. Hot water is more drastic and sodium carbonate solutions even more so. Excessive rinsing should be avoided because it removes wine that has soaked into the wood along with dissolved extractives. Some tartrates left in the barrel have the desirable effect of hastening tartrate deposition from new wine. Lees on the bottom of a barrel should be rinsed out before new wine is put in.

Before they are cleaned, barrels should be carefully inspected. If a barrel has been allowed to dry out too much it may have lost tightness and the metal hoops may need to be driven tighter. Barrels are also susceptible to attack by wood borers.

If a used barrel cannot be refilled with new wine within a day, the burning of a sulfur wick in it is recommended. Care should be taken to ensure that no particles of sulfur fall to the bottom; the result might be hydrogen sulfide production if a fermenting wine is ever placed in the barrel.

Used barrels can be stored dry or wet. Those stored dry should have sulfur wicks burned in them every week, and before reuse should be soaked with water for 12–24 hours and then thoroughly rinsed to remove excess sulfur dioxide. The danger with storing barrels dry is that the wood can shrink, causing leaks. Barrels stored wet should be filled with a liquid that will prevent microbial activity. A solution containing 1000 ppm of sulfur dioxide and 0.25% of citric acid is recommended. A disadvantage here is that extractives will be removed from the oak; the longer the storage the greater the loss. After prolonged storage, soaking overnight with a mild sodium carbonate solution at 120° F followed by thorough rinsing is recommended. Empty barrels, either dry or wet, should be stored for the least possible time.

If a barrel has developed a moldy smell after storage, soaking with a 1% sodium carbonate solution at 120° F can be tried. If this step does not help, the barrel should be discarded. In no case should a high-quality wine be placed in a suspect barrel; its soundness should be tested first by a period of storing a lesser wine. With spoilage inside a barrel, no amount of cure will equal good prevention.

Some winemakers in regions producing high-acid grapes, such as the eastern United States, have experienced a succession of years when desirable malolactic fermentations occurred in wines stored in oak barrels. Other winemakers in these regions have experienced undesirable lactic spoilage under similar circumstances. Winemakers who use oak cooperage cannot be neutral about malolactic fermentation. If they desire it, they should innoculate new wines with sufficient pure malolactic culture to ensure a clean fermentation and get these desirable strains of bacteria established in their barrels. If they do not desire it, they should rack wines early, store them at low temperatures, keep the sulfur dioxide moderately high, and perhaps add fumaric acid to the wines to ensure that the adventitious lactic bacteria in the winery do not get established.

Topping up Barrels

It is well known that storage containers must be kept as full as possible to prevent acetification of wines. Since the seepage of wine through pores in wood barrels lowers the level, winemakers traditionally fill (top up) barrels about once a week for the first few months and later about once a month. It is customary to turn barrels enough during the second and third years to cover the bung with wine and reduce air entry at this point, turning them upright for necessary toppings. Bungs have also been sealed with sealing wax. Wine loss is greater with smaller barrels, higher temperatures, and lower humidity.

In 1976, research by Richard Peterson at the Monterey Vineyard in California showed that a vacuum of up to ⅕ atmosphere developed in tightly bunged barrels in about 6 months. This vacuum varied greatly among barrels, depending on their tightness. The discovery led Peterson to suggest that most of the oxygen that gets into tight barrels enters during filling, topping, and emptying rather than during aging itself. Careful winemakers should check all barrels for tightness, either by inspecting for wet spots or by fitting them with vaccum gauges, as Peterson did. The sensory qualities of wines from individual barrels should be checked and recorded, and leaky barrels should be identified. The disposal of barrels giving poorer-than-average results should be considered.

Wood Extractives

Provided that barrel aging is not excessive, barrels contribute agreeable aroma and flavor components which can increase the complexity and quality of many types of wines. Some 5–10% of the total weight of dry oak consists of tannins and other substances that can be dissolved by wines.

The phenolic components of oak are largely nonflavonoids—in contrast to young wines, in which flavonoids predominate. Because of this difference, it is possible to identify wines that have been aged in wood by means of chemical tests. Vanillin, syringaldehyde, and related phenolics derived from lignin are among the desirable components contributed to wines by oak barrels. About 0.6–0.7% of the total weight of dry wood can be extracted initially. Reaction of lignin with water (hydrolysis) and oxidation apparently release more of these fragrant substances during aging.

Tests by V. L. Singleton at the University of California at Davis on 50-gallon American and European oak barrels show that in a new barrel about 3.8 grams of wood per liter of wine (½ oz per gallon) are extracted at a penetration of 0.5 mm in about 2 months. This is about

3–10 times the threshold taste level. In old barrels 6 mm of penetration is not unusual, which suggests that under ideal circumstances a barrel could be filled and emptied 100 times and give detectable flavor to all that wine.

If wine is aged in a new barrel for 2–3 years, about 70% of the total extractives will be removed—far too much for one wine. It is better if the first wine placed in a new barrel is removed as soon as sensory or chemical analysis indicates that the desired level of oak extractives has been achieved. As extractives are removed, a barrel can hold a wine longer. The most sensible procedure is to have two sets of barrels, a few new ones to impart extractives and more older ones for long-term aging. During its first year of use a barrel might be refilled with wine 3 or more times; by the end of the second or third year it would be one of the older barrels. After about 7 years barrels no longer provide useful levels of extractives.

Some winemakers attempt to save money by purchasing used barrels, from either wineries or distilleries. The loss of extractives should be considered when setting a value on used barrels. It should also be recognized that the liquids barrels previously held can contribute an undesirable flavor. Some eastern U.S. and Canadian wineries switched from labrusca wines to hybrid varieties and got a labrusca flavor in their new wines. Treatments to rinse out the old flavors will further deplete any remaining extractives in the wood. In California, however, some firms are equipped to remove wood from the inside of used barrels, thereby exposing fresh wood. This procedure can be done twice and by increasing available extractives can justify its cost.

In cooperage, the surface from which wines can gain extractives increases as the cube root of the volume. This means that increasing the volume 1000-fold increases the surface area only tenfold. In a 50-gallon barrel there is approximately 3.75 in^2 of surface per gallon of wine. A 500-gallon container has a little less than half that surface per gallon, and it takes more than twice as long to obtain the same level of extractives. A 5-gallon container has more than twice the surface area per gallon, but almost 5 times as many containers are needed to get the same amount of extractives into a wine in the same time. Given these facts, it is easy to see why most barrels used for aging table wines are close to 50 gallons in size.

European oak contributes more extract and tannin to wines than American and on average about ⅓ more flavor. American and European oaks have sufficiently different extractives so that experienced tasters can often tell which was used.

Opinion differs as to how much extractive is desirable in white wines, but the amount is perhaps ¼ that desired in red or dessert wines. Some experts feel that if the aroma and flavor of oak can be clearly detected in a white wine there is too much. Not all producers or consumers agree, however, and numerous examples of dry white wines with oaky character are on the market.

Oak Chips and Extracts

Both commercial and amateur winemakers frequently use oak chips (granular oak) to supplement or replace barrel extractives. Home winemakers, wishing to avoid the problems associated with barrels, can attempt to duplicate barrel aging by using oak chips and the carefully controlled introduction of air to wines in inert containers. Though such treatment may be less satisfactory than barrel aging, it is often better than no wood aging at all.

Oak chips yield extractives rather completely within a week but further contact can increase extractives by allowing hydrolysis of wood lignins. From ½ to 1 ounce of oak chips per gallon noticeably improves red wine quality. Experienced tasters can detect a difference in sensory qualities when a wine is treated with 1–3 grams of oak chips per gallon. This minimum detectable amount of oak is a good point at which to start experimentation with white wines.

Oak extracts are an alternative to oak chips and are used by some commercial wineries. An extract can be prepared by combining oak chips with a small amount of 100-proof alcohol (vodka is acceptable). Kept in a sealed bottle, it will lose no detectable quality in 10 years. Varying amounts of such an extract added to a series of wine samples can be used in tests to determine the desirable amount of oak aroma and flavor in a wine.

9

Wine Blending

A large portion of the world's wines—both ordinary and fine—are blended. Skillful blending, one of the winemaker's most useful tools, is an attempt to make two plus two equal at least five by diluting defects while improving balance and complexity.

Winemakers' goals in blending include: (1) a wine better balanced or more attractive than the individual ingredients; (2) A standard wine constant in quality year after year; (3) a wine produced at lower cost than would otherwise be possible; (4) a new type of wine.

Types of blends include: (1) wines from two or more varieties of grapes; (2) wines from one grape variety grown in different locations; (3) wines from two or more vintages; (4) wines that have received different vinifications; (5) wines from various casks; (6) nongrape ingredients together or with grape wines. Blends can include more than one of these types; thus the possibilities are almost endless.

Blending Strategies

Practical wine blending must take into account regulations (for commercial winemakers), availability of blending stocks, consumer preferences, economic considerations, and the skill of the blender.

Blending for balance usually involves a compromise and the goal is to improve weak characteristics more than existing qualities are sacrificied. General guidelines for blending are as follows:

(1) Only sound wines should be blended (a wine with only a slight excess of volatile acidity is a possible exception, since when it is dilute enough it is not objectionable).

(2) Grapes grown in warmer regions will generally be higher in sugar (potential wine alcohol and body) than the same grapes grown in cooler regions, while being lower in acidity, color, fragrance, and flavor.

(3) Undesirably strong fragrances and flavors are often improved by being blended with more neutral ones or with dissimilar ones. Among the strongest fragrances and flavors are those possessed by the labrusca and muscadine grape varieties of the eastern United States and the muscats of Europe and California. Some people have learned to appreciate these strong and distinctive odors and flavors, but broader consumer acceptance is often gained by careful blending to tone them down. Blending two or more distinctive grape characters (e.g., labrusca and muscandine) tends to reduce the prominence of each.

(4) Blending dissimilar premium grape varieties (e.g., Riesling and Chardonnay) together can neutralize their individual character too much. If the goal of blending is to change color, acidity, or other secondary characteristics, it is best achieved by using similar but lesser grapes (e.g., Merlot with Cabernet Sauvignon, Sauvignon blanc with Chardonnay).

(5) Blending semi-distinctive grape varieties—including many of the French hybrids and California varieties such as Chenin blanc and Pinot blanc—frequently gives more interesting wines than the individual pure varietal wines.

(6) Blending together wines from neutral grapes (e.g., Thompson seedless) is useful only in preparing preblends, since by themselves they usually have no special wine quality. (A preblend is a mixture of wines that will later become part of a final blend.) The main value of neutral grapes lies in their low cost and availability.

(7) The color and style of the wines blended should generally be similar (e.g., fruity red wines should be blended with other fruity reds), but there are exceptions to this rule. When one is dealing with the assertive fragrances of the labrusca varieties, the species character overrides individual varietal character, and reds such as Concord and whites such as Niagara can often be blended together without producing a strangely flavored wine. Experienced blenders

frequently blend some fruity wine with some that is less fruity to fine-tune the flavor.

(8) Blending wines produced by different vinification treatments is often an easier way to achieve a desired effect than trying to stop a particular vinification step at a specific point. For example, it is extremely difficult to predict when a Chardonnay aged in new oak barrels will reach the desired level of wood flavor. It is usually less trouble for the winemaker to allow some of the wine to pick up excess oak character and then blend that with other batches having less.

(9) Blends should be made with a specific purpose in mind, and no more ingredients should be used than the blender requires or has the skill to handle. Excessive blending can reduce any wine to mediocrity.

Planning the Blend

When planning a blend a winemaker should keep in mind consumer preferences, characteristics of available blending stocks, and the ways in which wines can be blended.

One can infer the preferences of consumers by noting the commercial wines that appeal to them. The limits of acceptability (such as amount of sweetness) should be determined when possible. Preferences vary with consumer sophistication, local food types, and fads. Novice consumers often prefer uncomplex wines with a pronounced aroma or flavor. In any locale, the introduction of unfamiliar wine types will be met with resistance.

Grape and other fruit compositions vary with variety, cultural practices, harvest year, and time of harvest, and winemakers should find out as much as possible about these variations. Depending on conditions, for example, a grape variety such as Baco noir can vary from 14° Brix and over 2% acid to 22° Brix and under 1% acid in vineyards in the same area. Some grapes such as Vignoles change noticeably in character as they ripen while other varieties do not.

The general strategy in wine blending is to identify the strengths and weaknesses of available stocks. In most cases one wine is chosen as the primary stock and others are tentatively identified as secondary blending stocks. The primary and secondary stocks may themselves be blends (referred to at this stage as preblends).

A preblend is prepared when no one wine has the characteristics desired in the backbone of a blend or in a secondary blending stock.

Perhaps a wine chosen as the primary stock has too strong a flavor. Other available blending stocks may have certain unbalanced secondary characteristics (such as excessive acidity or low body) that a preblend of several of them can balance out. Then the blender will be faced with only one problem to solve: too strong a flavor in the primary wine.

It is impossible to give a useful listing of all grape varieties that can be used in blends to accomplish a given purpose. The best guide here may be the composition of successful commercial blends, which are sometimes made known on back labels or discussed in consumer wine magazines, and winery newsletters. Commercial blends that have achieved consumer acceptance represent feasible solutions to particular blending problems. There are multivintage blends (wines from one grape variety grown in several years), multilocation blends (wines from one grape variety grown in several locations), multivariety blends (wines from several varieties), and multiblends (combinations of the above).

Once acceptable wine stocks have been identified, the winemaker should become even more familiar with each one. Sensory evaluation, supplemented where needed by physical and chemical analyses, is required. Analyses can be done by wine labs, other winemakers, or experienced friends if the winemaker cannot perform them. The number of analyses required depends on the goal of the blending program: if the goal is to reduce excessive acidity and no more, perhaps only the acidities of the blending stocks need be measured. But if a winemaker is attempting to duplicate a successful complex blend from a previous year, many analyses may be necessary.

The wine blender should know what can be accomplished with various types of blending. Multivintage blending is usually done either to raise the quality of wine from a poor year or to combine fresh and aged characteristics in a single wine. Multilocation blends are usually made to adjust subtle characteristics such as body or a particular shade of fragrance or flavor. Generally only large wineries, with multiple sources of grapes and sophisticated blenders, get involved in multilocational blending, but small winemakers purchasing grapes from several sources can also do so.

Multivarietal blends are more useful than others because there is usually more difference between varieties than between vintages or locations. This type of blending is used to stretch expensive varieties, to make a better balanced or more complex wine, or to create a style of wine that would be impossible with just one variety. Essentially all proprietary and generic commercial wines are multivarietal blends, whose creators aim for a distinctive style and for continuity in character from year to year. An example of multivarietal blending is found in the wines of Bordeaux where the grapes are good in most years and blending is used to balance tannins, acidity, and color. Multivarietal, multivintage, and multilocation blends are common for nonvintage French Champagnes and Portugese port wines whose producers use all possible means to achieve year-to-year consistency.

The time when wines are blended influences the end result. Some winemakers are experienced enough to be able to blend grapes in the fermenter. This method is practical, of course, only when grapes ripen at the same time or can be held in cold storage until needed. Some larger eastern U.S. wine producers have found that when Concord grapes are fermented with California grapes (or juice) the resulting wine is better than if it were combined after each type of grape was fermented separately. It takes a great deal of skill and experience to predict what a single wine will eventually taste like from sampling a fermentation or a very immature wine, and the added complications of trying to assess a blend during fermentation are beyond the abilities of most beginning wine blenders.

Blending Tactics

Once a specific blending goal has been set and a base and secondary stocks selected, blending can begin. The process should be carried out under conditions conducive to accurate evaluation. It is best to work in a quiet room with light-colored walls and a minimum of distractions. Wine cellers, which are often not well lit and frequently have distracting odors, are not good places to blend; a laboratory or kitchen is more suitable.

One or more tasters should be assembled to evaluate the trial blends. Large wineries use several staff members and may bring in outside experts for an important blending session. Besides competence in evaluating wines, the qualifications of the tasters should include familiarity with the types of wines to be blended, and knowledge of and ability to reflect the tastes of consumers. It is a mistake for winemakers—even home winemakers—to rely totally on their own personal preferences when blending a wine. Involving family members or friends can keep a home winemaker from producing wines that only he or she can appreciate. Sometimes winemakers build up a tolerance for certain flaws in their own wines—high acidity, excess sulfur dioxide, mercaptan smells, high volatile acidity—and do not realize that others find these things objectionable. Consumers are more definite about their dislikes than their likes, and a blender should learn the former. Among the factors influencing consumers' odor and flavor preferences, age and wine-tasting experience are very important.

Tasters evaluating blends should have a clean work area, water to rinse the mouth, a receptacle to spit wines into, and clean, code-numbered glasses filled by a nontaster (to avoid prejudice). Several blends should be presented at a time so that they can be compared: for example, 3:1, 1:1, and 1:3 blends of the primary wine and a secondary blending stock. With practice the wine blender will have a good idea of suitable starting proportions and can save time at this step.

The blends should be tasted along with the unblended components, and each glass of wine should be carefully evaluated and scored on points of interest. After the evaluation, when the identity of each sample is made known, tasters can decide if any of the blends is superior to the unblended components and what further improvements might be made.

If a blend is poorer than its components, tasters' notes should shed light on why this is so. Some wines just do not go well together, and discovering these incompatibilities is part of learning to be a competent blender. Certain good wines cancel the desirable qualities of others, and some excellent wine stocks may have to be rejected because they do not fit into the particular blend being worked on at the moment.

If a blend is superior to its components but falls short of the blending goal, a second round of blending should be done, with the proportions of blending stocks adjusted or new stocks introduced. If possible, the blend judged best should be tasted blind against a wine that approximates the goal—perhaps a commercial wine or a previous successful blend. Using a benchmark helps to validate or correct the tasters' memory of their goal. If a goal proves beyond reach with available materials, then it must be changed. Accurate information on the qualities of available stocks, combined with experience, can reduce the number of false starts to a minimum.

Wine blending, like any learned skill, must be practiced frequently if it is to be maintained in top form. Those aspiring to be competent blenders should practice throughout the year, not just during the winemaking season. Wine-tasting events, meals where several wines are served, and practice blending sessions are all valuable training.

Careful notes taken during practice sessions can prove invaluable later.

Many people have slight "blindnesses" in wine evaluation; odor and taste thresholds vary, and often a person may be inexperienced with certain wine characteristics. Beginning wine blenders should test their own evaluations against those of more experienced people to upgrade skills and to learn to identify more nuances.

One useful tool for planning wine blends is the Pearson square. Often recommended for use with alcohol fortification to produce dessert wines, it can actually be used for any wine constituent that can be measured. The square appears in this form:

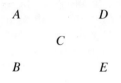

A and B represent the amount of some component in two different wines (A being the larger), C represents the amount of that component desired in a blend, and D and E represent the proportion of wines A and B in the blend. ($D = C - B$ and $E = A - C$.) An example of blending to achieve desired acidity will serve as an illustration.

Given two wines, one with 0.95% acid and one with 0.55% acid, place the value 0.95 in position A and the value 0.55 in position B. If a desired blend should have 0.70% acidity, this value is placed at position C. D is then the difference between 0.70 and 0.55 (0.15) and E is the difference between 0.95 and 0.70 (0.25). Putting these values in place gives the completed square:

.95   .15

.70

.55   .25

To get the desired 0.70% acid in the blend, the proportion of the wine with 0.95% acid to that with 0.55% acid should be 0.15 to 0.25 (or 3 to 5). The same problem can be worked out with algebra by solving the equation:

$$0.95\,D + 0.55\,E = 0.70\,(D + E)$$

With blends of several wines it is possible to optimize as many quality factors as there are wines in the blend (e.g., with three wines one can try to optimize color, acidity, and body at the same time by adjusting the quantity of each wine in the blend), provided that suitable wines are available. The mathematical technique used is called linear programming. Explaining its details is beyond the scope of this book, but interested readers can refer to the mathematical literature. (Blends of up to 4 wines can be handled on a programmable calculator and more complicated blends worked out on a personal computer.)

After a desirable wine blend has been identified, one should prepare a slightly larger quantity (perhaps a few gallons) and store this for a month or two before reevaluating it. Instabilities in the blend (such as can occur when a wine that has undergone a malolactic fermentation is blended with one that has not) can be detected, as well as possible improvements as the various fragrances and flavors merge. If this intermediate scale blend is stable and holds or improves its quality, the final blend can then be prepared.

Wine blending is practiced in every wine region. A few representative examples will illustrate the scope of problems that blending can correct. Here we will focus on one problem at a time, though winemakers usually try to solve as many problems as possible when they blend.

Blending for Color

Most color problems involve red wines. Grenache, for example, which is important in the warmer districts of southern Europe, California, and Australia, is a premium variety when not overcropped, but its color is so weak that it must usually be blended with Carignane or other dark grapes to make a normal-colored red wine.

During blending to correct color, a winemaker must guard against unbalancing other components. Cabernet franc, for example, is used in Bordeaux to add color, but because it also adds acidity it seldom makes up more than a third of a blend. A winemaker can add color to a red wine without upsetting its balance by blending in a few percent of a wine made from a teinturier variety. If properly ripe, these are so strongly colored that they can be used as dyes. Alicante Bouschet and Colobel (Seibel 8357) are hybrid teinturiers, and Tar Heel is a rotundafolia variety. (They have been used in California, the northeast, and the southeast respectively.) Such grapes often make up less than 5% of a blend.

One reason for blending to correct a color problem is to compensate for color changes caused by unstable grape pigments. Pinot noir,

Baco noir, and essentially all dark muscadine varieties have unstable pigments that tend to brown and fade fairly rapidly during aging. Winemakers using such varieties might consider blending in a tein-turier to prolong appealing color.

Blending for Fragrance

Blending with the aim of diluting an overly fragrant wine is practiced commercially in the eastern United States, where producers blend fairly neutral California wines with assertive native American varieties, such as Concord, to make their product more acceptable to consumers. Some blending for fragrance is similar to adding a tein-turier for color. In France, small amounts of Muscadelle are sometimes blended with the wines of Graves, Barsac, and Sauternes to impart a floral character.

The fragrances of wines are so complex (hundreds of components, many below threshold detection levels, have been identified) that we really don't know why some wines smell better than others. When one particular group of components is present in enough quantity, we can identify fragrances that remind us of fruits, flowers, vegetation, and other things. The very best wines have subtle and interesting combinations of these fragrance groups. Cabernet Sauvignon, for example, reminds some people of berries, green peppers, and cigar smoke. Although such a rough description may not appeal, those smelling the wine are aware of many fleeting impressions that combine to give pleasure.

While most premium grape varieties have pleasing mixtures of fragrances, however, other varieties present a one-dimensional odor. Certain French hybrids, for example, have a strong weedy smell; others have a strong berry smell. When these individual varieties are made into pure varietal wines, they are generally disappointing, and the best winemakers have learned how to combine two or more of them to produce a more interesting result.

Blending for Taste

Blending to produce a balanced taste is usually easier than blending for fragrance because laboratory analyses can guide the wine blender.

Only three basic tastes are found in most wines: sour, sweet, and bitter.

Grape acidity varies with variety, cultural practices, and the temperature of the growing season. In cool climates or seasons the acids stay at a high level; in hot climates or seasons they decrease. Winemakers in California's hot Central Valley often blend in grapes from the cooler coastal regions to raise acidity whereas major wineries in the northeastern United States purchase Central Valley wines to lower the acidity of local ones. Since few of the world's grape- growing areas have an ideal climate, the blending of wines from hot and cool regions is widely practiced.

Grape sweetness depends on sunshine and the length of the growing season. Blending is seldom used to reduce the sugar content of table wines. Districts with long, sunny growing seasons usually specialize in sweet dessert wines and have no need to lower their sugar content. In many cooler districts, however, blending to adjust sweetness is widely practiced. In Germany, for example, the cool climate gives wines of fairly high acidity which benefit from having some sweetness. Many German winemakers ferment the main bulk of their wines to dryness, then blend in a "sweet reserve" of slightly fermented grape juice. Winemakers with modern equipment can avoid this type of blending by stopping fermentation at the desired sugar level and stabilizing wines with a sterile filtration.

Wine bitterness is mostly due to tannins and varies with grape variety and skin contact time. The classic example of adjusting tannin level by blending is found in Bordeaux. Cabernet Sauvignon is known for its fine varietal fragrance and flavor, but it gives tannic and slow maturing wines. Merlot is less tannic and faster maturing, but often less interesting. Many of the finest Bordeaux wines are a careful blend of the two (along with Cabernet franc and Petit Verdot in some cases). In California, pure Cabernet Sauvignon wines were made for many years, but some winemakers now blend Merlot with them and a few blend Cabernet Sauvignon with their Merlot wines.

Blending for Alcohol and Body

Alcohol content is sometimes adjusted by blending wines from hot climates (with high alcohol) into those from cool climates (with acceptable color, fragrance, flavor, and acidity).

In producing fine table wines in cooler districts, winemakers usually achieve an adequate alcohol level by reducing the size of the grape crop so that there are adequate leaf surfaces to ripen the grapes. In countries such as France, where there is a very large demand for ordinary wines, growers frequently overcrop their vines and get wines thin and low in alcohol, so producers of ordinary French wines have long used the wines of the Midi, Algeria, and Italy to increase the alcohol in their blends. Up to a point, higher alcohol makes a better wine by increasing body, reducing chances of spoilage, and adding a slight sweetness to balance an otherwise acidic or harsh taste. With modern techniques there is less excuse than there once was for producing wines high in alcohol, and the continued sale of these in France owes more to cultural than to technical factors.

The body of a wine is related to its alcohol content because alcohol increases viscosity and feel of fulness in the mouth, but other factors (e.g., carbohydrates, higher alcohols, phenolics) also influence body. A winemaker may be able to blend wines from several locations to achieve desired body without knowing exactly why the mixture works. Different grape varieties seem to vary in the amount of body they provide in wines. In the eastern United States, Chambourcin, a French hybrid, has found favor with some wine producers because it enhances the body of blends. This grape has been commercially available only since 1963 and is already one of the most propagated hybrids in France, where presumably it is also favored for its body.

Blending for Wood Extract

When wines are aged in small barrels the amount of wood character they pick up varies greatly with the age of the barrel. In a new barrel a wine may take on a noticeable oak flavor in one month, while the same wine in an old barrel may not pick up much in two years. Winemakers can age the bulk of their wine in older barrels and store only a small portion in new ones, then blend the wines to get the desired degree of oak character. As a substitute for new barrels, granular oak can be used to flavor a small batch of wine.

When the wood character in barreled wine is adequate, a winemaker usually blends together wine from all the barrels before bottling, to prevent bottle-to-bottle variation. In France this is called "equalizing the vintage." Inexperienced commercial winemakers in the United States have been known to bottle wines of one vintage out of individual barrels over a period of years, with the result that a consumer can never be sure that the wine in two apparently identical bottles will be similar.

Blending for Oxidation and Maturity

In the traditional Spanish sherry method of fractional blending (the solera system), young wines are blended with older, more oxidized ones. This process provides year-to-year consistency and gives a complex wine containing both fresh and aged components. French Champagne producers often blend wines of several vintages together to achieve the maturity desired in their nonvintage wines. A few California wine producers do multivintage blending to achieve complexity.

In selected cases, blending wines having slight volatile acidity with wines without much fragrance can enhance the character of a blend and simulate fruitiness.

Other Blending

Not only can blending correct deficiencies, it may also improve the quality of wines without noticeable defects. Research in California has shown that mixing two commercial wines of similar quality rarely reduces quality and often improves it, probably because of increased complexity. This suggests that producers of unblended vintaged varietal wines may not be achieving the highest possible quality. After all, if simply mixing two wines in equal proportions can raise quality on an average, skillful blending should be able to do much more.

In the Champagne district of France, blending is very sophisticated and wines are blended not only for the usual factors of balance and complexity but also to get a flavor that will peak at the time the wines will be consumed. In most vintaged Champagne wines, which are intended to age for 8–10 years, only Pinot noir and Chardonnay grapes are used, with the former predominating. But nonvintage Champagne wines, meant to be consumed earlier, include up to ⅓ Pinot Meunier, a variety that matures more rapidly in the bottle than the other two and peaks in about three years, roughly the right time. In the taste of such a wine, Pinot Meunier provides the first impression, while Pinot noir and Chardonnay give a longer-lasting impression. In a vintage Champagne Pinot noir would so overwhelm the Pinot Meunier that there is little reason to include the latter.

Producers in the sherry as in the Champagne district, usually strive for year-to-year consistency. These two areas pretty much represent the southern and northern limits of premium wine production in Europe. Spain has few years that are exceptionally bad; Champagne has few that are exceptionally good. In both regions winemakers rely heavily on blending to achieve the most interesting wines possible.

Blending to minimize year-to-year variations and surprises is the norm in most of the world's wine districts. The Bordeaux region of France is the model for the vintage-year gamble—a gamble for both producers and consumers, since both invest their money without knowing exactly what they are going to get. The element of risk here lends excitement and glamor to the wine scene, and many people are enthralled by it. There is little reason to doubt, however, that judicious multivintage, multivineyard, and multivariety blending could raise the average quality of Bordeaux wines.

In the United States and other countries, high-quality grape varieties are sometimes diluted with more neutral and less expensive ones to lower costs. The range in U.S. grape prices from the top varieties to the bottom is often greater than a factor of ten. There is some evidence to suggest that diluting a very characteristic variety does not dilute odor and taste impressions as much as might be expected. With careful blending, the increased complexity of a wine can sometimes offset the loss of varietal character. (Some producers of varietally diluted wines fail to pass savings along to consumers and frequently put them into advertising instead. This practice has given some people the false impression that wine blending invariably cheats the consumer.)

Sometimes premium grape varieties are blended into lesser grape wines without the fact being indicated to the consumer. This has happened in the eastern United States when wine producers have had a small amount of a premium variety, such as Chardonnay, but not enough for a separate bottling. It happened in California when the switch in consumer preferences to white wines left varieties such as Cabernet Sauvignon without a home. Producers frequently do not reveal the presence of premium varieties in blends in the hope that they can withdraw those varieties at a later date and put them to better use.

Blending also provides a method for winemakers to produce new types of wines. Many pop wines are blends of grapes and other fruits, and many proprietary wines are distinctive blends outside of traditional types. Some winemakers who hesitate to produce new types of wines may miss good bets. When wines made by the *maceration carbonique* process were investigated in California and the eastern United States, for example, limited consumer tests indicated that they were no better than wines produced by traditional vinification methods. But in the author's experience, blends of *macération carbonique* wines with traditional wines of the same type are often better than either by itself, suggesting that careful blending could produce a new wine type that consumers would accept.

The advantages of blending are not limited to grape wines. Many fruit wines benefit from blending; examples include apple and cherry wines, which often are best when they combine sweet and tart varieties. In areas of the world where it is too cold to grow grapes, enterprising winemakers have attempted to duplicate common types of grape wines by using fruits and berries. A Danish supplier of wine-making kits offers a concentrated blend of apples, elderberries, and gooseberries that reasonably simulates a red grape wine. Although such combinations will never truly duplicate their models, they show that creative blending can provide interesting new wine types.

10

White Table Wines

From the time of Napoleon to nearly the present, red wines were more popular than whites (often by 4 to 1) because they were more suitable for drinking at meals when people did hard physical work and ate hearty foods. In the past few years white wines have become more popular in the United States and elsewhere, perhaps partly because of a shift to lighter diets.

White table wines of the past generally had moderate levels of alcohol and flavor and were drunk at meals that featured fish and white meats. White table wines are now also consumed before or between meals and at social occasions, where they have partly displaced cocktails and aperitif wines. Both lighter and heavier styles of white wine have recently appeared.

Making a good white wine requires finesse, and until recently many small winemakers found whites difficult to make. With advances in technical knowledge, however, most winemakers now find them as easy to produce as reds.

White Wines from Grape Concentrates

Many amateurs start their winemaking with grape concentrates. If one is careful and uses those of high quality, it is possible to produce an acceptable white wine. Because concentrates rapidly lose quality in storage, those less than 3 months old are preferable. It is almost impossible, however, to produce a delicate, fruity white wine from concentrates since much of the grape aroma is lost during concentration. Home winemakers will often be better satisfied if the concentrate they choose is made from grapes whose quality is not dependent on a delicate aroma (e.g., native American or muscat varieties). Many concentrates give wines with a noticeable baked or caramelized flavor, a defect that the pungent labrusca or muscat aromas can hide.

Many recipes provided with grape concentrates yield a semi-sweet wine stabilized with sorbates. Since some people can taste them, the use of sorbates makes for lower wine quality. Because they are not completely reliable in preventing refermentation of sweet wines, such wines should be consumed soon after production.

Despite the disadvantages of grape concentrates, they have a legitimate place in amateur winemaking. Most are balanced to give wines with proper alcohol, acidity, and body, and additive packets for further balancing are supplied with some concentrates. Directions are usually clear and prevent major errors. Success with concentrates often leads amateur winemakers on to other techniques. Wines from concentrates can also serve as blending stocks for adjusting the acidity, body, or other characteristics of wines from fresh grapes. The fairly neutral character of many white wines from concentrates often makes them suitable bases for sherry, vermouth, and other flavored wines.

EXAMPLE 10A

Producing a wine from grape concentrate

A 46-fluid-oz can of concentrate is dissolved in a gallon of hot water, and 6½ lb of cane sugar in 2 gallons of hot water. These solutions are mixed, diluted to 5 U.S. gallons with cool water, placed in a 7-gallon plastic fermenter. When the mixture has cooled to 70–75° F, a packet of wine yeast is added. The fermenter should be covered with a top or with plastic wrap tied in place.

When the fermentation has subsided, the wine is siphoned into a glass carboy and a fermentation lock attached. The wine is racked in 3 weeks and again in 3 months, with the carboy kept full in the meantime.

When the wine is clear, it can be bottled. If the winemaker so desires, it may be sweetened with cane sugar and stabilized with sorbate to retard fermentation. It should be aged for a few months in a cool, dark place.

Standard-Quality White Wines from Grapes

California has a Mediterranean climate with mild wet winters and hot dry summers, ideal for grape growing. Wine grapes at harvest in California often have 20–25° Brix, 0.5–0.9% acid, and 3.2–3.8 pH. Much of the rest of North America has a four-season climate, which is less favorable for growing wine grapes. Especially north of 40° latitude, rains during harvest and early frosts often force harvesting when grapes are underripe and high in acidity. Eastern grapes at harvest often have 12–20° Brix, 0.8–1.7% acid, and 2.8–3.4 pH. Some hybrid and vinifera grapes in select eastern locations achieve the ripeness of California grapes, but this is relatively rare.

Producing standard-quality white wines in California requires grapes with normal sugar content and techniques that preserve grape qualities. With many eastern grapes, proper maturity is often more related to the desired wine flavor than to sugar content.

Whether winemakers grow their own grapes or purchase them, they should monitor ripeness to choose harvest dates. Most small-scale winemakers use hydrometers to measure grape sugar. Since different grape bunches reach maturity at different times, it is necessary to sample grapes throughout the block of vines to be harvested for a particular wine. With a suitable sample of grapes in hand, a winemaker crushes them, filters or settles the juice to remove grape particles (which cause high readings), places a hydrometer in a cylinder filled with the juice, and takes a reading of the sugar content from the stem of the floating hydrometer. Hydrometer readings change with temperature, so if the juice temperature does not match that of the hydrometer calibration a temperature correction should be made. A reasonable correction can be figured by subtracting 0.03 °Brix for each °F below the calibration temperature or adding 0.03 °Brix for each °F above that temperature. A possible source of error: Because hand squeezing does not crush unripe grapes to the same extent that pressing does, grape ripeness may be overestimated by 1° Brix or more, misleading the winemaker into harvesting grapes too soon. A refractometer, which can measure the sugar in one drop of juice, is small enough to carry in a pocket. Although this instrument is expensive compared to a hydrometer, it is more useful; it can be carried

Hydrometer

around a vineyard to sample berries and is not influenced by grape pulp in the juice.

Just before harvest a winemaker should measure the total acidity of grapes and test for pectins. When grapes are too low in acids, adding a fruit acid (such as tartaric acid) at pressing will usually give a better wine than adding the acid later.

If a pectin test is positive—fairly rare in grapes—the winemaker should add a pectic enzyme at crushing time. Commercial pectic enzymes vary in strength, and winemakers should follow the instructions provided with them. Enzymes usually need to work for 1–4 hours before pressing to reduce pectins adequately and increase juice yield. They can increase pressing efficiency, especially when grapes have slippery skins (as some eastern varieties do) or a slimy feel. Modern pure pectic enzymes are safe to use even on the better grape varieties destined for quality wines.

After harvest, grapes should usually be pressed as soon as possible to minimize quality losses. An exception might be made for sound grapes harvested during the hottest part of a day, which the winemaker may want to let cool overnight before crushing. Grapes that begin fermentation at too high a temperature can lose much of their fragrance during the early stages of the process.

It is always desirable to add sulfur dioxide to crushed white grapes to retard oxidation and browning and to control undesirable microorganisms. The SO_2 dosage depends on grape condition, must pH, and other factors but will usually be 50–100 ppm.

In eastern wineries, where amelioration (sugar and/or water addi-

Refractometer

tion) is very common, some wineries get fermentations started with large amounts of yeast starter, then add the required amount of sugar when the Brix has fallen to about 8°. This tends to increase fermentation speed.

EXAMPLE 10B
Producing a standard-quality white wine

Four grape lugs (150 lb) of Seyval blanc (a French-American hybrid) grapes are harvested at 21° Brix and 0.95% total acidity. The grapes are crushed (with 50 ppm of sulfur dioxide added) and pressed in a basket press with the stems left in as a press aid. After the first pressing, another 25 ppm of sulfur dioxide is added to the pulp, which is stirred up and pressed again. A third pressing follows the same procedure.

The grape juice (about 11 gallons) is placed in glass carboys and allowed to settle overnight. The next day the clear juice is racked from the pulp into containers, which are filled about ⅚ full. Additional juice is obtained by combining the pulp from all containers and resettling it. A yeast culture is prepared from 2 packets (10 grams) of dried Champagne yeast mixed with several ounces of water at 100° F and allowed to stand for 15–30 minutes. Eight grams of diammonium phosphate are dissolved in water and mixed with the grape juice, which is then innoculated with the yeast.

Fermentation begins in 24–48 hours. After roughly a week at 65–70° the wine has a Brix of about 0° and is allowed to ferment to dryness, which takes about another week. The end of the fermentation is ascertained with Dextrocheck® tablets. When the wine is dry (showing less than 0.2% residual sugar and no signs of fermentation), it is racked into fresh containers that are filled as full as possible and closed with air locks; 25 ppm of sulfur dioxide is added.

The wine is treated with 2 grams of bentonite per gallon to clarify it and cooled to about 45° F, if possible. After a week it is racked from the bentonite lees and returned to ambient temperature. The total acidity is now about 0.83%, and the volume is about 10 gallons.

When the outside temperature falls to about 30° F, enough potassium bicarbonate is added to reduce the acidity to 0.65% (3.4 grams per gallon will reduce the acidity by 0.1%) and the wine moved to an unheated building. Cream of tartar (1 gram per gallon) is mixed in. As the wine is cold-stabilized at 25–35° F, the precipitated potassium bitartrate crystals are stirred up and mixed with it every few days. After about 2 weeks the wine is brought inside and immediately racked, avoiding excessive air contact. Another 25 ppm of sulfur dioxide is added.

After three months the wine is again racked, away from a slight sediment. It is filtered through a 0.5–0.65-micron filter, 25 ppm of sulfur sulfur dioxide is added, and the wine is bottled. It has about 11.5% alcohol, medium body, and suitable acidity. In 6 months to a year it is ready to drink but should remain in good condition for 3 years or more.

General Quality Factors in White Wine Production

While large wineries enjoy the advantages of sophisticated equipment and highly trained people, small wineries and home winemakers can spend many hours nursing small batches of wine through all the steps that high quality demands.

Some winemakers feel that they can produce the highest-quality wines only if they grow their own grapes, but few people possess the skills to be both expert grower and expert winemaker. Many commercial wineries have made a name for quality using purchased grapes (or even purchased wines), and many amateur winemakers have won awards with wines made from purchased grapes.

Evidence suggests that winemakers who strive for the highest quality should concentrate on 2–5 types of wines and avoid spreading themselves too thin. Producing only one type is usually advisable only with something like a port wine from the hot regions of California, where climate variation is minimal. Grapes in cooler regions show greater annual variations, and winemakers can reduce their risks by making wines from several grapes that ripen at various times.

Along with harvesting at the optimum time, discarding rotten berries, and ensuring that grapes are not too hot when ready for crushing, one of the most important quality factors is pressing technique. When grapes are pressed, unless something is added to provide channels for juice to escape, pressing efficiency is low. When crushing grapes, some winemakers leave the stems in so they will act as a press aid. Stems add a little tannin to white musts and can aid clarification. Many commercial winemakers fear that the added tannin will be too much for their delicate white wines, however, and prefer other press aids such as rice hulls.

Several types of presses are in use. The vertical basket press is most popular with amateur winemakers, while small wineries favor the horizontal basket press. Juice that drains from a press with a minimum of pressure is often known as "free-run" juice, while that obtained during heavy pressing is called "press juice." After grapes are crushed, a winemaker has a choice of methods for juice removal: (1) putting the must directly into a press, (2) first dejuicing the must and then pressing, or (3) holding the must in contact with skins for some time before pressing. The choice affects the free-run juice yield, the percentage of solids in the juice, and other factors of juice composition.

Pressing immediately after crushing is simple, requires less equipment than other methods, and is the method usually used by home winemakers and small wineries for whom speed and efficiency of pressing are not critical. It is also the method of choice for grapes in poor condition because it minimizes off aromas and oxidation.

Commercial wineries sometimes remove juice before pressing to reduce the number of press loads. The use of moving screens often gives higher solids content than a stationary screen.

Juice composition changes during pressing. Free-run and lightly pressed juice usually constitutes 50–60% of the total and contains 1–4% solids. As pressing pressure increases, phenolics and pH increase and acidity drops. Total juice yield in typical commercial pressings is 160–180 gallons per ton though home winemakers usually get less.

Free-run and press fractions of juice are sometimes kept separate and may be reblended later, though many winemakers do not separate them. Juice is often chilled and allowed to settle overnight, then racked away from grape sediment. Some wineries use a centrifuge in place of settling, but others believe that settling works as well and improves quality. In any event, there is some consensus that removal of solids before fermentation is essential to quality in white vinifera wine. With native grape varieties used in the east, settling is uncommon because with many of these varieties the pulp does not separate enough to make the step worthwhile. Some winemakers even feel that settling reduces wine quality with these grapes.

Must held for a few hours or days before pressing is sometimes kept in contact with skins. This method, which yields 70–120 gallons per ton of free-run juice with relatively low suspended solids (0.5–3%), is not employed in Europe but is widely used on the north coast of California, especially with heavier-style wines. It can change the juice composition (lowering acidity and increasing phenolics and potassium) depending on the grapes, length of skin contact time, and temperature (higher temperatures speed the effects). The °Brix may increase during skin contact if grapes are overripe and some are shriveled or botrytised. Some research indicates that skin contact up to 16 hours (a moderately long time) can increase aromas in Chardonnay with no significant increase in bitterness or astringency. Other research shows, however, that moderate amounts of skin contact time lead to decreased sensory quality. Clearly, the issue of the benefits of skin contact for white wines has not yet been resolved.

Winemakers who favor skin contact believe it increases varietal flavor. Others think that varietal character can better be gained by picking grapes later. Some winemakers adjust skin contact time depending on flavor intensity in each variety (e.g., they give Sylvaner, which has less flavor than Gewurztraminer, more time).

During pressing and settling, juice is in contact with air. Oxidation of white grape juice causes browning and is accelerated by natural grape enzymes, such as polyphenoloxidase. Laccase, an enzyme found in grapes affected by botrytis, also contributes to browning. Sulfur dioxide reduces browning by deactivating these enzymes and because of its antioxidant effect. Certain grape varieties brown rapidly, but the reducing atmosphere during fermentation largely reverses the process. If air exposure does not adversely affect flavors, slight browning prior to fermentation is tolerable.

Some protection against the deleterious effects of air is given by sulfur dioxide. The amount of SO_2 added at crushing varies with the winemaker but the majority prefer 50–100 ppm. (Moldy or hot grapes, or those low in acidity need more SO_2). Wines made with up to 50 ppm of SO_2 are better than wines made without any, but higher levels (e.g., 200 ppm) can interfere with fermentation and increase acetaldehyde. Grapes blanketed with carbon dioxide need less SO_2.

If not enough sulfur dioxide has been used at crushing, more can be added to the juice before fermentation. SO_2 should not be put in during fermentation because it increases acetaldehyde. Commercial winemakers usually try to maintain 25–30 ppm free SO_2 after fermentation until bottling.

The desired acidity of white grape juice is 0.7–0.95%. When juice is low in acidity, acids can be added before fermentation or wines can be adjusted later. If acid is to be added, many smaller wineries prefer tartaric acid (a natural component of grapes) while larger ones prefer citric acid (which minimizes cold-stability problems). Malic acid may be added when a malolactic fermentation is desired.

Some grape musts, especially those from underripe grapes, benefit from added sugar. Grape concentrates provide sugar and other qualities. Although regulations permit it, few if any premium wine producers in California add concentrates to wines, either during fermentation or later to sweeten them. Home winemakers in cooler areas of the United States and Canada may find that high-quality, fresh grape concentrates can sometimes add body, balance, and complexity to wines.

Fining can be done before fermentation (gelatin reduces tannins in press juice and bentonite reduces proteins), but it is more efficient afterward. Since suspended particles promote fermentation, overfined white grape juice may be difficult to ferment.

A variety of yeasts are used for white wines. Four commonly used by north coast California winemakers are: Montrachet (University of California at Davis #522), Champagne (UCD #505), Steinberg (Geisenheim), and French White (Pasteur Institute, Paris). Steinberg is losing popularity and Epernay 2 may replace it. Most commercial wineries use about 2% by volume of an actively growing yeast culture to innoculate grape juice. Fresh juice can be innoculated with rapidly fermenting juice, but there is some risk of contamination.

Some American winemakers prefer to ferment in stainless steel while others claim that fermenting in barrels creates a better flavor. Controlling temperature is easier with barrels than with large unrefrigerated tanks but refrigerated tanks are better still. Barrels, on the other hand, are hard to clean and impossible to sterilize. Wine picks up oak flavor more rapidly during fermentation than during aging, probably because of the stirring action of the bubbles. Available evidence suggests that barrel fermentation may temporarily change the character of a wine but that after some oak aging such wines are not necessarily better or worse than wines fermented in stainless steel or glass and may indeed be indistinguishable.

Winemakers disagree on the optimum temperature for fermenting white wines. Evidence suggests that wines that owe their charm to a fresh, fruity character ought to be fermented at 45–55° F (or sometimes lower), rather than at room temperature, to preserve fruitiness. This procedure is generally followed in Germany and takes 2–8 weeks or more. The fermentation bouquets formed are sometimes short-lived and such wines reach peak quality quite quickly (sometimes in 6 months). Storing them at low temperatures prolongs their fruitiness. White wines destined for aging, which derive their quality from factors other than fruitiness, may benefit from fermentation at 60°F or higher. (White Burgundy wines are fermented at about 68° F). Higher temperatures seem to permit the extraction of more flavor from grapes while allowing more odor loss. Some U.S. producers ferment Chardonnay at about 50° F until it reaches 10–12° Brix, then rack the wine into barrels to finish the fermention at ambient temperatures.

Commercial wineries cool fermenting wines by a variety of methods, usually involving refrigeration. Small wineries and home winemakers often place fermentation vessels in a cool room. For small batches, home winemakers can use a spare refrigerator set at the highest temperature setting. A new trend is to control the fermentation rate rather than let it start fast and soon slow as it naturally does. Winemakers sometimes adjust temperatures to decrease sugar by 1–1.5% per day down to about 8%, then readjust the temperature to about 60° F to finish the fermentation.

Winemakers use various methods to adjust SO_2 after fermentation and racking: adding fixed amounts of the compound, adjusting free SO_2, or adjusting total SO_2. At this stage free SO_2 should be about 25 ppm and total SO_2 should be no higher than 100–150 ppm.

A malolactic fermentation in white wines is fairly rare in California but more common for commercial eastern U.S. wines and French Chardonnays. If one wants to encourage it, the general method is to keep free SO_2 low, use a moderately high (65–70° F) fermentation temperature, allow the lees to remain longer than usual, and innoculate with a pure malolactic bacteria culture.

In cellar practices with white wines, the main precaution is to minimize oxygen pickup. Some commercial wineries use oxygen meters to monitor wines during cellar treatments. Inert gases (nitrogen or carbon dioxide) can be used to blanket partly full containers of wine to protect them from oxygen.

The lees of high-quality white wines are removed as soon as possible after fermentation and the wines are clarified. The simplest clarification method is to rack the wine several times over a period of months. (Wineries with centrifuges often use them at this stage.) A diatomaceous earth (D.E.) filtration is often used by small wineries and some home winemakers after preliminary settling has taken place. Filtration with pads or cartridges requires a cleaner wine to start with but often gives better clarity than a D.E. filtration.

Fining can be done at several points during cellar treatment: in the juice, after fermentation, and just before bottling. Fining juice, usually less efficient than fining wine, is sometimes done to reduce tannins in hard-pressed juice fractions so that they can be combined with lighter-pressed juice for fermentation. A protein such as gelatin is used for this type of fining. Bentonite fining is usually done after fermentation to reduce proteins and clarify the wine. Sparkolloid® and gelatin-tannin are also used for fining white wines. Fining before bottling is sometimes done to improve a wine's sensory qualities, for example, to reduce astringency.

Sophisticated winemakers aim with fining, filtering, and centrifuging to remove from wine components that would lower quality if left in. A few white wines do not require fining, but most need filtering to be of high quality and acceptable to consumers. It is generally agreed that clarification treatments should be held to the minimum necessary for acceptable clarity.

Clarification is often combined with a stabilization step to minimize handling of the wine. A typical procedure is to rack the wine off yeast lees, bentonite fine for heat stability, chill for cold stability, then rack the wine again and use a D.E. filter to remove remaining particles. Many commercial wineries use a membrane filtration before bottling.

Aging improves wines because some of the constituents chemically react and are converted to new compounds which have better fragrance and flavor. Wines designed to have a fresh, fruity flavor are usually aged for only a few months or a year in stainless steel, glass, or other inert containers. Some grapes such as Chardonnay, Sauvignon blanc, Chenin blanc, and Pinot blanc can be made into either fresh, fruity wines or aged, richer, more complex ones. If the latter are desired, these wines are usually aged in oak barrels for 2–12 months, depending on the amount of oak flavor wanted and the age of the barrels. The quality of the wine one starts with will determine the amount of aging it can take. Thinner, more delicate wines can be easily overpowered by oak flavors. European oak has generally been found to give milder oak flavors than American and is more often used with white wines. Wines mature faster in barrels at higher temperatures; cellar temperatures for barrel aging are usually 60–68° F.

Most commercial wineries bottle white wines 6–10 months after the harvest, except for a few that can benefit from longer aging. Home winemakers have more freedom (not having to worry about work or cash flow problems) in selecting bottling times but should bottle as soon as white wines are ready.

The general commercial practice is to adjust the free SO_2 to 20–30 ppm at the time of bottling. Home winemakers without facilities for measuring SO_2 can usually be safe by adding about 30 ppm of SO_2 at bottling.

Commercial wineries that can afford the expense use bottle fillers that employ either a vacuum or an inert gas to reduce oxygen in the bottle headspace. Fastidious home winemakers sometimes blow nitrogen gas through a wine for about 10 minutes before bottling and may also flush bottles with nitrogen. Although the author has found noticeable quality improvements in wines blown with nitrogen within a few weeks after bottling, he does not know if this advantage remains after several months.

Some European white wines owe a portion of their charm to residual carbon dioxide present at the time of bottling. There is some reason to believe that the use of carbon dioxide to flush containers during the later stages of winemaking enhances white wine quality by reducing the loss of residual carbon dioxide and the pickup of oxygen. (Commercial North American winemakers have to be careful that the carbon dioxide content of their wines is not too high, or they will be taxed more heavily.)

Premium White Wines—Fresh, Lighter Style

Winemakers have learned in recent years that the fruity qualities of many white grapes can best be preserved when fermentation occurs in the range of 40–55° F. Cool fermentation, the practice for white wines in Germany, is becoming popular in North America.

A variety of yeasts are used to enhance the fruity character of white wines, though it is not always certain whether the yeast strains actually increase wine fragrances directly or merely slow down fermentation to make possible the retention of natural fragrance. Slower-growing yeasts offer a benefit if one is producing wines with residual sugar since fermentation is easier to stop. Vigorous yeasts tend to give hydrogen sulfide, a problem that slow-growing strains minimize. The main dangers in a slow fermentation, especially with a weak yeast, are that it may fail to start properly or may stick before it has proceeded as far as desired.

Some winemakers find drawbacks to stopping a fermenting wine to get sweetness—such as the need to rack or filter it many times—and think that adding a little "sweet reserve" of grape juice is easier.

For years the best-selling foreign wines in the United States have had alcohol strengths mostly below the minimum 10% that California required for white wines. It is estimated that in 1980 about ⅔ of the most popular imported wines were in the 7–8% alcohol range. A December 1979 change in state regulations permitted California winemakers to produce wines with as little as 7% alcohol, the federal minimum. A number of American wineries are now producing "light" table wines that are dry or almost dry and are 7–9½% alcohol. A second class of wines low in alcohol is being produced medium-dry or sweet and labeled "soft." At least some consumers prefer these wines with food, believing that a high alcohol level detracts from a grape's varietal character.

EXAMPLE 10C

Producing a light-style Riesling wine

Riesling grapes are harvested in a cool climate at 17° Brix and 1.2% acidity. The procedure outlined in Example 10B is followed, with modification. The settled grape juice is innoculated with a liquid culture built up from Epernay 2 yeast, and when fermentation starts a few days later the temperature is slowly reduced to about 45° F. Fermentation lasts 4–7 weeks. When taste tests of the wine show that it has the desired residual sugar (about 2%), it is chilled to 30° F to stop fermentation, racked, and treated with 50 ppm of sulfur dioxide. The wine is kept cool and a week later racked again, or alternatively, filtered through a 0.65-micron or smaller-pore filter. If it is not filtered, the wine will have to be kept cold and racked several times to eliminate yeasts.

Prior to cold stabilization, the wine is treated with Acidex® or a similar product to reduce the acidity to about 0.85%. At about 4 months of age, it is put through a 0.5-micron filter, and 50 ppm of sulfur dioxide added before bottling. The wine at this point has about 8% alcohol, 0.8% acid, and 2% residual sugar. It should be consumed within about 6 months after bottling.

Methods for producing light wines vary. Some winemakers seek out grapes from cool growing areas, such as California's Monterey

County, where varietal character can be achieved without high sugar content. In the Mosel area of Germany, grapes are often harvested at 15–17° Brix and still have ample flavor and aroma. Other winemakers have been experimenting with underripe grapes—fermenting them to dryness, partly dealcoholizing the wine under vacuum, and then blending in small amounts of wines with high intensity flavor and aroma. It is too early to tell if these "light" and "soft" wines will permanently attract consumers, but they offer the added attraction of having fewer than the usual calories.

Many amateur winemakers have had access only to grapes from cool areas and have faithfully followed directions—in winemaking books and other sources of information—to add sugar to grape must to obtain 22° Brix. If the grapes they use are of good quality, many of these winemakers may be pleasantly surprised by the wines they can produce by leaving out the sugar.

Premium White Wines—Aged, Heavier Style

Many winemakers in California and some in other parts of the United States have attempted to duplicate the white wines of Burgundy. To do this they usually try to harvest grapes as ripe as possible, ferment at moderate temperatures, and age the wines in oak barrels for several months and then in bottles for a year or more before releasing them. Wines produced in this way often owe more of their character to their treatment and the development of bottle bouquet than to the original fresh fruitiness of the grapes. But the use of premium grapes is essential.

The following example indicates how even premium grapes sometimes need special treatment to make a fine white wine. Readers are not expected to follow all details of this example but may learn about some pitfalls and techniques from it.

EXAMPLE 10D
Producing a heavy-style Chardonnay wine

A 30-gallon container of frozen California Chardonnay grape juice (from grapes harvested at 24° Brix, 1.02% total acidity, and 3.05 pH) is thawed during 5 days at a cellar temperature of 60° F. The juice is siphoned and dipped from the container, leaving a considerable amount of tartrates behind, to yield 26 gallons of juice with an acidity of 0.76% and a pH of 3.20. The juice is transferred to 5-gallon glass carboys and allowed to settle while 0.5 gallon of a Champagne yeast culture is built up from 10 grams of dried yeast. After a day the settled juice is racked from grape particles, ⅔ g/gal of diammonium phosphate is added, and the active yeast is added to the juice. Fermentation starts within a day or so and is completed in about 4 weeks at 60° F.

The wine is racked, and after 25 ppm of SO_2 is added, fined with 2 g/gal of bentonite and allowed to settle for 2 weeks before racking. After several months it is racked again, and another 25 ppm of SO_2 is added. Portions of the wine in small containers that have darkened are treated with 2 g/gal of PolyClar AT®. The acidity of the dry wine is 0.82%, and its electrical conductivity is 2500 micromhos/cm. (Conductivity results from salts in the wine, potassium bitartrate being a major one. Measuring conductivity is a convenient way to follow the progress of cold stabilization.) The tartrate content of the wine is measured to be about 0.2%—too low to permit further acid reduction by carbonate addition. Tartaric acid is added to raise the acidity to 1%, and the wine is treated with enough Acidex® to reduce the total acidity to 0.7%. After the wine is cold-stabilized at 25° F for 2 weeks in a freezer controlled with an external thermostat, it is racked; its conductivity is then 2000 micromhos/cm and its acidity is 0.68%. (Ideally, the wine should have a conductivity of about 1600 micromhos/cm, but with the very low tartaric acid content, 2000 micromhos/cm is probably adequate to ensure cold stability.)

The wine is stored in 5-gal glass carboys for 4 months with 1/2 oz of oak chips in each carboy. It is then filtered through a 0.5-micron cartridge, and 30 ppm of SO_2 is added. The various carboys of wine have distinctly different characters, so they are blended before bottling. The newly bottled wine is disappointing for the first year, but after a 1200-mile trip by moving van in the summer, it develops a definite and desirable French white Burgundy character.

In the 1970s California winemakers explored the limits of heavy-style white wine production. Grapes were allowed to remain on vines until their sugar reached far above normal Brix levels. Some grapes were grown in areas where rainfall was low enough to stress the vines and desiccate the berries. Crushed grapes were allowed to remain in contact with their skins for up to 2 days before pressing. Grapes were fermented in barrels and later given extensive barrel aging in new oak. The resulting wines often had 14% or higher alcohol and frequently had intense flavors and oak character. Some of them won prizes at wine competitions and found a following amoung small groups of consumers. This extreme style of winemaking decreased, however, as observers noted that highly alcoholic and intensely flavored white wines were caricatures of table wines and had little use in normal circumstances. Winemakers now tend to stay within traditional imits in making heavy-style white wines.

White Wines from Native Grapes

In the eastern United States and Canada white wines are frequently made from native grape varieties. Labrusca varieties, including Niagara and Delaware; muscadine varieties, including Scuppernong and several newer ones such as Magnolia, Welder, and Miss Blanc.

Most of the principles of white wine production outlined above apply to working with native varieties, but there are exceptions. Native eastern varieties have very strong flavors and fragrances and often benefit from being picked before they are completely ripe. Because the grapes are relatively low in sugar when picked, they are often ameliorated with sugar. Labrusca varieties often have high acids and are ameliorated with water.

These grapes have tough skins and require more pressing pressure than vinifera varieties. Some commercial winemakers using muscadine varieties believe quality is improved by limiting juice yield to only 100 gallons per ton. Many of these seem to benefit from blending with more neutral grape varieties.

Few labrusca or muscadine white wines seem to improve with age, and aging in oak is almost unknown. By and large one has to learn to appreciate these wines, and a taste for them is generally limited to people growing up in areas where they are the major wines produced. Winemakers in the southeastern United States are making strides in improving muscadine wine quality, however, and their wines are frequently in great local demand.

11

Red Table Wines

Throughout most of history red wines have been more popular than whites. Red wines go well with many meals and generally keep for several years. Two hundred years ago the reputation of fine red table wines was well established in France. With advances in technology, good reds are no longer limited to Europe. Modern winemakers everywhere have learned that they need to match good grape varieties to proper climates and soils, that they should use proper yeast and bacterial cultures, and that barrel aging is important in producing high-quality red wines.

Red Wines from Grape Concentrates

Producing a red wine from a grape concentrate is similar to producing a white wine. The major problem with concentrates here is that fermentation on the skins is not possible and wines therefore lack the color, flavor, and tannins of those from fresh grapes.

Probably the most successful red wines from concentrates are rosé table wines. A rosé wine is produced from red grapes when skin contact time is short, typically just a few hours. Many rosés do not owe their charm to distinctive grape varietal flavors or fragrances.

They have a pale red color and moderate fragrance, are well balanced, and are often slightly sweet. These are characteristics that a grape concentrate can provide, and rosé concentrates often yield a good rosé which can benefit from a year of bottle aging.

Standard-Quality Red Wines from Grapes

In traditional processes, red wines differ from whites principally because they pick up color, flavor, and tannins from contact with grape skins during fermentation. Pigments and tannins in skins are more soluble at higher temperatures and in alcoholic solutions. (Example 1A in Chapter 1 gives a simple red wine recipe.)

Fruity red wines, for quick consumption, are made much like white wines. In the eastern United States and Canada, Foch is one of several varieties that lend themselves to this style. Fruitiness is preserved by fermenting at moderate temperatures, pressing the pomace before fermentation is complete, avoiding a malolactic fermentation, and bottling as soon as the wine is clarified and stabilized (4–8 months after harvest). Fruity, unpretentious, dry red wines are common in Europe and elsewhere. Some are similar but off-dry or semi-sweet. This style of wine is suitable for everyday drinking and can be produced with minimum effort.

Home winemakers can make a light red wine by adding sugar water (about 2 lb of sugar per gallon) to pressed red grape pomace and conducting a second fermentation and pressing. The volume of such "press" wine should be limited to less than half the volume of the first wine, and the wine should be consumed quickly.

Hot-Pressed Wines

Large wineries in eastern North America frequently use hot pressing in place of fermentation on skins. In this technique grapes are crushed and stemmed, heated to about 140° F for a half hour or more, cooled, treated with pectic enzymes, then pressed with the help of rice hulls or other press aids. The juice is fermented and treated much like a white wine. Hot pressing has been most used with labrusca grapes, the skins of which are difficult to crush, and it results in adequate color while minimizing tannins. It is appropriate for Concord and other wines produced in a fruity, often sweet style without wood aging. With red varieties of more delicate flavor (e.g., French-American hybrids), heating may be limited to 120° F.

Since heating grapes releases pectins while reducing pectic enzymes, pectic enzymes need to be added before or during fermentation to clarify the juice. Heated musts tend to be less bright red than those fermented on skins, but proper hot pressing does not adversely affect flavor. (Hot-pressed grapes should not be fermented on skins, or the wine taste will be inferior.)

Some eastern U.S. winemakers have been experimenting with hot pressing of hybrid grapes, which are crushed, heated to about 180° F for a few minutes, then cooled and pressed. They claim that this technique gives fresh, fruity, light red wines with less harsh flavor than results from fermentation on the skins. Home winemakers who want to try this technique on a small scale can drop whole bunches of red grapes into a large pot of boiling water, take them out after a minute, drop them into a pot of ice water, then press them.

General Quality Factors in Red Wine Production

Quality factors for red wines are similar to those for whites. Yields of premium grapes (such as Cabernet Sauvignon and Pinot noir) average 1–4 tons per acre. Those of lesser varieties are sometimes quite a bit higher. A few winemakers in good climates keep crop levels below normal by cluster thinning or farming on dry hillsides, claiming that the intense varietal character is worth the added expense. Others feel that bringing out very intense varietal flavor from Cabernet Sauvignon and some other red grapes is not necessarily good and prefer grapes from level vineyards with moderate yields. Winemakers in less favorable climates often have to be satisfied with whatever yield they can get.

Most red grapes are harvested at 20–24° Brix, depending on the climate. Some winemakers allow red grapes to become extremely ripe to obtain very heavy dark red wines. Extreme examples (sometimes 15% or more alcohol) have been produced in California but have a limited market. In general, underripeness in red grapes is more of a negative quality factor than it is in whites. Relatively few North American growing areas outside California are capable of producing highest-quality red grapes, but reasonably good ones are grown in many parts of the United States and Canada.

Grapes are usually crushed thoroughly to expose both insides and outsides of skins to the fermenting juice for the best extraction of color and flavor. Some winemakers leave some stems in the crushed grapes to provide additional tannin, especially with a variety like Pinot noir, which has low skin tannin.

The amount of SO_2 added to crushed red grapes varies. Sound, moderately acid grapes—especially those destined for malolactic fermentation—may need less than 50 ppm. When grapes are in poor condition (e.g., moldy) or low in acid and no malolactic fermentation is wanted, 100–150 ppm of SO_2 may be used. In general, 50 ppm added at the crush helps reduce wild yeasts and bacteria and has some antioxidant effect—important when producing red wines in a fruity style.

In warm climates, winemakers sometimes add acids to crushed grapes deficient in them. (See the earlier discussion on white grapes.)

For red wines, fermentation precedes pressing. Most commercial winemakers avoid reliance on natural yeasts and innoculate red grape must with a pure yeast culture. A few small California winemakers have experimented with natural yeasts but do not seem very confident that the result is actually a better wine—at least not consistently. In areas outside California, where wine yeasts have not been spread by the return of pomace to vineyards, the use of wild yeasts seems much riskier and has little to recommend it.

When several varieties of red grapes are to be combined in one wine, it is often best to make the blend at the crush or soon after fermentation. Early blending, especially if the grapes differ considerably in flavor, often gives more pleasing results than blending finished wines.

Many small commercial wineries ferment red grapes in stainless steel vessels. Open-topped containers make for better cooling and probably for a more homogeneous temperature during fermentation than do closed vessels. Home winemakers usually find food-grade plastic containers easiest to handle and clean.

Most winemakers ferment red grapes at 70–85° F. Above 90° F, yeasts are weakened and begin to die. At 70–85° F, fermentation and color extraction progress faster than at lower temperatures while tannin extraction is not proportionally increased. Cooling during the most vigorous part of the fermentation is sometimes required. The winemaker's ideal is to extract from the grapes just those things that he wants, and temperature control is one method of balancing the color/tannin extraction ratio.

During fermentation, skins rise to form a cap on top of the wine. If

Punching down the "cap" during a red wine fermentation

the cap is allowed to remain out of the wine too long, vinegar or other spoilage can occur and color and flavor extraction decrease. Small wineries therefore either: (1) punch the cap down into the wine by hand, (2) pump the wine over the cap 2–4 times a day, or (3) use a grid below the surface to keep the skins constantly submerged. A sub-merged-cap fermentation, however, may overextract skins and result in too pungent a wine. Punching down the cap, the only method used by home winemakers, affords good control of the extraction process.

Most color is extracted early while tannin extraction continues later into the fermentation. Long skin contact increases tannin but not color, and excessive contact may actually reduce color. In a warm season with very ripe grapes, a shorter period of skin contact should be permitted than in cool seasons. A few wineries use skin contact times of 2–3 weeks before pressing, but the wines require many years to smooth out. Most home winemakers will get more pleasing results with 2–7 days of skin contact time.

Skin contact ends with pressing. The crucial factors in pressing appear to be how hard the grapes are pressed and which press frac-tions are combined. Most small wineries producing premium reds use relatively gentle pressing and limit juice yields to around 150–160 gallons per ton, about the yield that a home winemaker can get with a hand-operated basket press. By blending moderately pressed juice with free-run juice, the winemaker can increase color, tannins, com-plexity, and richness. In some cases it may be advisable to ferment separate press fractions and blend them after they are aged.

Many winemakers encourage a malolactic fermentation in red wines by innoculating them with malolactic bacteria before or during the fermentation. Some wineries have indigenous malolactic bacteria, but natural malolactic fermentations can cause off odors and flavors. Because the bacteria need temperatures above 65° F to grow, winemakers may have to warm wines. After a malolactic fermentation has ended, the wine should be racked, aerated, and treated with 50–75 ppm of SO_2. In some warm climates grapes contain so little malic acid that malolactic fermentation is very mild or does not occur at all. High-alcohol wines from very ripe grapes may likewise not go through a malolactic fermentation even if one is desired.

Most red wines are cloudy after either kind of fermentation. Some wineries clarify wines by racking them every 3 months for several years, but one fining and one or two filtrations can accomplish more clarification in a shorter time. It may be desirable to clarify wines before putting them into barrels to minimize sediment formation in the barrels. Some winemakers rarely fine or filter their red wines, the dark color masks problems of haze or sediment. Available information suggests, however, that winemakers who avoid filtering because they believe it takes character out of a red wine are misinformed as to what desirable character is.

Fining is not always necessary with red wines since tannins from skins combine with grape proteins to form a built-in fining system. The most popular fining agent for red wines is probably gelatin, and its use is usual, after malolactic fermentations. Properly fined wines will give little or no sediment in the bottle during normal storage. Winemakers should experiment to find the minimum amount of fining agent to reach the desired result. As little as ⅛–¼ g/gal of gelatin is used, but more may be needed to reduce tannins and soften the flavor of certain red wines. Some winemakers believe egg whites to be more gentle than gelatin; one or two whites per barrel of wine is about the average. Other winemakers use bentonite before gelatin fining. Chilling combined with fining often increases the fining's effectiveness.

Most winemakers rack red wines frequently just after fermentation (every few weeks or every month) and less frequently (1–4 times per year) during aging. Four times may be too many. Those who rack wines out of barrels as often as every three months and wash the barrels out are often the same people who reject fining and filtering. This extra handling of wines is of questionable value. Racking with a siphon can separate wine from lees better and result in more clarity than racking with a pump.

Many winemakers do not cold-stabilize red wines, apparently believing that they will never be subjected to cool temperatures. Most red wines not cold-stabilized, however, will eventually deposit tartrates in the bottle. Home winemakers may choose to ignore these deposits, but the settling out of tartrates during cold stabilization can remove bitter-tasting oxidized pigment particles and both clarify a wine and improve quality. Heat stabilization of red wines, via bentonite fining, is rarely necessary since most proteins (which cause heat instability) are precipitated by the tannins. Some rosés benefit from a bentonite fining.

Most commercial wineries age premium red wines in 50–60-gallon oak barrels. Home winemakers, of course, often use smaller barrels. Some wineries use new barrels each year, but many phase in new barrels and retire old ones on a continuing basis.

The desired length of time for barrel aging varies with grape variety, wine style, and barrel size and age. When used continuously, a barrel is depleted of wood extractives in about 7 years. Depending on the wine, the aging period in the barrels may be 1–2 years or just a few months. Because new barrels quickly impart oak flavor, some winemakers use them for only a portion of their wines, put the rest in older barrels or other containers, and later blend the various batches to get the amount of oak flavor they want. There are noticeable differences between red wines aged in French (and other European) oak and American oak barrels. In most cases which is used comes down to the winemaker's personal preference.

The advantages of blending red wines are being more appreciated by U.S. winemakers. Cabernet Sauvignon is frequently improved by the addition of Merlot and vice versa. Some believe that a wine loses varietal character when the proportion of the main grape drops below ⅔, but high-quality red wines are frequently composed of 3 or more grapes. Many wines made by amateurs can benefit from judicious blending.

Before bottling, many commercial wineries pump wines from individual barrels into a tank to mix them, ensuring uniformity from bottle to bottle. But before blending various barrels, one should segregate any that show off smells or atypical flavors. In many cases, after suitable treatment, they can be blended with the main lot. If a bacterial infection has started but has not caused irreversible odor or flavor changes, some winemakers add all the SO_2 destined for the entire batch to the infected barrels to kill the bacteria, then blend this wine with the rest to dilute the SO_2 to the desired level.

Not all red wines benefit from barrel aging. Fruity types—including rosés—are generally treated much like fruity whites and bottled within 4–8 months after harvest. Some of these will benefit from a year or more of bottle aging; others will reach peak quality sooner and should be consumed within a year.

Just before bottling, many commercial winemakers give wines a final filtration, through a regular fine filter (0.45–0.65 micron) or sometimes an absolute membrane filter (0.2–0.45 micron). This step eliminates bacteria that might cause problems in the bottle and removes tiny oxidized pigment particles that would otherwise throw a sediment within a few years. Home winemakers and some small wineries often skip this final filtration.

To decrease further any chances of bacterial action in the bottle many commercial wineries adjust the free SO_2 level to about 25 ppm at the time of bottling. Most homemade wines would benefit from the same procedure, but because free SO_2 is difficult to measure home winemakers usually add about 40 ppm of SO_2 and leave it at that.

Rosé Wines

In recent years rosé wines have become more popular; by 1979 rosé table wine shipments from California exceeded red. The light flavor of rosé wines and the fact that they are often served chilled means that they are used like whites. A well-made rosé is light and delicate but has enough flavor to go with many foods.

Rosé wines are usually red wines that have been removed from contact with the skins before much color, flavor, and tannin have been extracted from them. Skin contact time is usually just a few hours, though it depends on grape variety and ripeness. In the classic method, free-run juice is drawn off, after a brief fermentation, to make a rosé wine, then the remaining juice is fermented on the skins to make a red. Rosé wines can also be made from ripe, highly colored red grapes by pressing before fermentation. Such wines resemble white wines more than reds and are usually handled like whites.

Rosé wines made from a single grape variety can have a distinctive, but muted, varietal character. Some California wineries have gained a reputation for varietal rosés made from premium grapes (e.g., Cabernet Sauvignon), but only a few have specialized in rosés of the classic French Tavel type, which are made from Grenache grapes generally grown on dry hillsides, where the yield is low.

Rosé wines can also be produced by blending red and white wines. Some winemakers believe that the result is a fresher, fruitier wine, but these are essentially just tinted whites; such blends, frequently made by large eastern wineries, have no distinctive varietal character. They are finished sweet or semi-sweet, and many are quite pleasant.

When feasible, it is desirable to store rosé wines at lower than usual cellar temperatures to help maintain their fresh, fruity character as long as possible.

EXAMPLE 11A
Producing a rosé wine

Grapes that have proved suitable for rosé wine production—Gamay, Grenache, Pinot noir, Zinfandel, Cascade (Seibel 13053), and other varieties—can be used. After they are crushed and fermentation begins, samples of the fermenting wine are drawn off every hour or so until the color reaches almost the desired level. The free-run juice is then separated and fermented. After a little more time, somewhat darker juice can be withdrawn and fermented. Remaining juice can be fermented on the skins to obtain a normal red wine.

The lighter-colored wines should be handled like white wines, kept cool and given enough SO_2 so that a malolactic fermentation is avoided. When they have been clarified and stabilized, they can be blended to achieve the best color, body, and flavor. The rosé wine should be bottled within 4–8 months and consumed within a year or two, while it is still fruity.

A slightly sweet rosé can be produced if one adds cane sugar or grape concentrate and stabilizes the wine before bottling.

Nouveau Red Wines

"Nouveau" red wines, with a fresh, fruity, and frequently raw taste, are very popular in parts of France and are exported to other countries. Some appear on the market in Paris a few months after harvest.

The classic method of producing nouveau wines in Beaujolais involves the *macération carbonique*, which gives wines with a distinctive fragrance and flavor. Whole grapes are placed in a container filled with carbon dioxide and permitted to stand for a week or more while natural enzymes soften them until they ooze juice. Fermentation starts from the natural yeasts on the skins, and the grapes are pressed when it is well underway. When the wine has fermented dry, it is usually held at 65° F or above to encourage malolactic fermentation, which generally follows very quickly. When that ends (usually within a few weeks) the wine is racked, aerated, and treated with SO_2. It is then clarified by fining and/or filtration and bottled. Stabilization against heat and cold is usually not necessary since these wines are meant for very quick consumption.

In recent years a few North American wine researchers and wineries have attempted to duplicate the French carbonic maceration process, but have reported poor results, particularly high volatile acidity. In the author's opinion, success with this process depends on several factors. Only sound grapes without any mold should be used. Carbon dioxide dissolves in water, and a container filled with grapes can absorb enough carbon dioxide to create a partial vacuum, sucking air into the container. The container should therefore be flushed with carbon dioxide several times until there is no tendency to absorb more. Although French grapes are usually coated with wine yeasts, many of the wild yeasts found in North American vineyards are not desirable wine types. It is a good practice, accordingly, to put a pure strain of wine yeast in the container to ensure a sound fermentation. And since the carbonic maceration process tends to promote a malolactic fermentation, it makes sense to innoculate with a desirable malolactic culture at the beginning.

EXAMPLE 11B
Producing a red wine via carbonic maceration

DeChaunac, Pinot noir, or other red grapes with a good sugar/acid balance are carefully destemmed by hand and placed in a glass, stainless steel, or plastic container. (For large batches, destemming may not be practical and whole bunches can be used. Though the stems increase tannin content, the amount may not be excessive. If it is, fining with gelatin will reduce tannin.) A pure yeast culture and a malolactic culture are prepared and sprayed on the grapes before the container is closed.

The container is flushed with carbon dioxide gas, then sealed with an air lock. When the air lock indicates a partial vacuum (water is sucked back into the container), flushing with carbon dioxide is repeated until no further vacuum is formed. (Alternatively, the addition of a small amount of actively fermenting grape juice to the container may eliminate the need for carbon dioxide.)

The container is stored at a moderate temperature (60–70° F) until bubbles in the air lock indicate that fermentation is underway. It usually starts in 10 days to 3 weeks. The grapes are then removed from the container, pressed, and treated much like a white wine. Fermentation should be finished at a relatively low temperature to protect fruity aromas.

In many cases a malolactic fermentation is completed soon after the yeast fermentation and can be verified by paper chromatography. If this process does not get underway on its own, the wine can be reinnoculated with a malolactic culture.

When the wine has finished the malolactic fermentation, it can be fined (with gelatin if it is too tannic), filtered, treated with about 50 ppm of SO_2, and bottled. It should be consumed within a few months while its nouveau character is still apparent.

Carbonic maceration probably will not become very popular in North America; it requires extra effort and expense and some consumers do not like the distinctive flavors. But wines of this type add interesting flavor notes when blended with normally produced wines from similar grapes. Many of the bland red wines produced in North America, without distinctive varietal character or wood aging, could be candidates for this type of blending. (Winemakers should keep in mind that blending a wine that has undergone a malolactic fermentation with one that has not will often cause the process to begin again, possibly in the bottle.)

Premium Red Wines—Aged, Heavier Style

The main requirement for a premium red wine of heavier style is fully ripened premium grapes. Nothing that can be done in the winery will substitute for top-quality grapes. Given good grapes, the winemaker needs to decide how much color, flavor, and tannin is wanted in the wine.

EXAMPLE 11C

Producing a Cabernet Sauvignon wine

Cabernet Sauvignon grapes are harvested at 23° Brix and 1% total acidity from a relatively cool growing area. They are crushed and stemmed, treated with 40 ppm of SO_2, placed in an open-top fermenter, and innoculated a few hours later with a Montrachet, Pasteur Red, or Champagne yeast culture. If previous experience indicates that hydrogen sulfide is likely to be produced with the yeast being used, about ⅔ gram per gallon of diammonium phosphate should be added at the crush.

When fermentation begins, the cap is punched down 2–3 times a day, and the wine is smelled daily to detect any off odors. When fermentation subsides and the Brix drops to 5°, the grapes are pressed. The free-run and lightly pressed fractions are combined, and heavily pressed fractions are put into separate containers. All wine fractions are innoculated with a pure malolactic bacteria culture and fermented to dryness. The end of the yeast fermentation is determined with Dextrocheck® test tablets, and the progress of the malolactic fermentation is followed with paper chromatography.

When malolactic fermentation is complete, the wine is racked. If it tastes too tannic it is fined with enough gelatin (determined by lab trials) to reduce the tannins, then racked again and treated with about 50 ppm of sulfur dioxide. Trial blends with the most heavily pressed wine may indicate that some of that fraction should be blended in for additional flavor or body. After several months, when outside temperatures have fallen, the wine is cold-stabilized for several weeks, after which it is racked again and treated with another 25 ppm of sulfur dioxide. The acidity at this point should be about 0.65%; if it is too low or high, adjustments can be made.

The wine is then placed in oak barrels, or else in glass or other inert containers with one or more oz of oak chips per gallon added. When tasting indicates that the desired amount of oak flavor has been extracted, the wine is racked and filtered through a 0.65-micron filter. If older barrels are used, the wine may remain in them for up to 2 years.

The wine is then critically smelled and tasted and, if it seems desirable, a portion of Merlot or other grape wine of a compatible type may be blended in to provide balance and complexity. A final 40 ppm of sulfur dioxide is added and the wine is bottled. It should be ready to drink in 1–3 years but may improve with longer bottle aging.

Red Wines from Native Grapes

Dark-skinned labrusca and muscadine varieties have thick skins and generally need to be treated like the corresponding white varieties described in Chapter 10.

Red muscadine wines are usually less successful than whites. The red wines have strong and sometimes unpleasant flavors and unstable pigments that fade and turn brown after a year or so. Tar Heel is a dark muscadine that has been used to increase the color in blends. Researchers in Florida, Mississippi, and elsewhere continue to experiment with these varieties in hopes of improving palatability, but generally admit that white muscadines are superior for winemaking.

12

Sweet Table Wines

Many unfortified wines, including both table and dessert types, contain unfermented (residual) sugar. Most consumers cannot distinguish wines with 0.5% sugar from completely dry wines but when sugar is at 1% almost everyone can taste sweetness. Worldwide, most consumers prefer dry wines with meals and sweet ones after or between meals but many American consumers prefer sweet wines any time. A study in California found that about half the population preferred sweet wines to dry. Recognizing this, some wineries market essentially the same wine in both dry and moderately sweet styles. Thus, many wines containing up to 3% sugar are sold as table wines (e.g., California sauterne). Wines above 3% sugar, however, are usually considered dessert or aperitif wines (e.g., California "chateau" sauterne). French Sauternes, which usually contain more than 10% sugar, are invariably used as dessert wines.

Sugar adds sweetness and body to wines and balances their acid taste, but tends to mask other flavors. Wine connoisseurs are aware of the masking effect and some view all inexpensive sweet wines with suspicion. On the other hand, it is known that certain grapes (e.g., Riesling and Chenin blanc) give fruitier, more appealing wines when they contain some residual sugar.

One way to classify sweet wines is by the way residual sugar gets into the wine:

(1) By winemaker intervention to stop fermentation before all sugar is gone. This method is widely used around the world, especially with grapes that are quite ripe.

(2) By using grapes dehydrated by a special mold to give an extremely high sugar content. Fermentation usually stops by itself with high residual sugar in the wine. Some luscious dessert wines are made this way in Europe and California.

(3) By adding grape juice, or partly fermented grape juice, to dry wine. This is often the method of choice in France, Germany, and North America, especially with grapes that are not very ripe and that tend to ferment dry very easily.

Grapes for Sweet Table Wines

French Sauternes are made from about ¼ Sauvignon blanc grapes and ¾ Semillion, blended at the crush. The same varieties are considered to make the best sweet California table wines. Chenin blanc and Muscat varieties are also used. Chenin blanc wines retain more fruitiness when fermentation is stopped before dryness.

Some expensive and highly prized sweet wines, including German beerenausleses, French Sauternes, and Hungarian Tokays, are produced from grapes whose flavors have been concentrated by a special type of mold, *Botrytis cinerea*. This fungus removes water from the grapes, increases glycerine content and oxidizing enzymes, and decreases total acidity. It generally reduces malic and tartaric acids about equally. The °Brix of affected grapes can be 26 to over 50. The fungus requires a very short period of moist weather to become established, but a significant dry period must follow. When wet or foggy weather follows the initial botrytis infection the grapes usually rot. The effect of botrytis can be controlled by timing the harvest. Grape yields are often no more than 0.5–1 ton per acre and juice yields only 80–120 gallons per ton. Not all varieties benefit from botrytis dehydration; some undergo an excessive loss of fruitiness while others show very little increase in sugar.

The production of wines from botrytis-infected grapes in the United States has recently increased but is still small. French Sauternes-type wines are few because conditions for a natural botrytis infection rarely exist. Wines with synthetically induced botrytis have been made in California. The method is to arrange ripe grapes on shallow trays, spray them with botrytis spores, then cover them to keep the humidity as high as possible. After 24 hours the cover is removed and the grapes maintained in lower relative humidity at around 70° F for about two weeks. The result is very high quality sweet table wines with good botrytis flavor.

Sweet wines without botrytis are made from late-harvested grapes or sweetened dry wines. Undercropped vineyards can produce premium varieties with high sugar content. Grapes harvested late should not be permitted to raisin on the vine because that gives undesirable flavors and too dark a color in the wine. It is desirable to pick the ripest grapes for sweet wine production and then harvest the remaining ones for dry wines. Removing raisined or rotten berries by hand improves quality.

Production of Sweet Wines by Incomplete Fermentation

Wineries with access to very ripe grapes prefer musts with high sugar content for making the sweeter types of wine. Musts below 24° Brix used to make sweet table wines by incomplete fermentation give lower alcohol and extract than normal dry table wines. A total acidity of 0.8–0.9% in musts is sometimes needed to give sweet wines with a good sugar/acid balance.

Winemakers who prefer to make sweet table wines by stopping the fermentation short of dryness (rather than by completing it and adding sugar) believe that this method gives them more natural wine of higher quality and with a lower alcohol level.

Grapes for sweet wines are handled much like those for dry. They are rarely fermented on the skins although with shriveled grapes some skin contact might make for greater sugar extraction. Crushing semi-dried grapes is much more difficult, and one must usually slow the pressing to get the maximum juice.

Musts from overripe grapes are often high in suspended solids, and the wines may be difficult to clarify. It is desirable to settle these musts at a cool temperature before fermentation, thus removing some nutrients and wild yeasts, and gaining better control of the fermentation. Pectic enzymes may be used to advantage during settling. Up to 100–150 ppm of sulfur dioxide can be added.

A pure yeast culture is preferable to wild yeasts, and should be added 1–4 hours after the sulfur dioxide or following racking if the must is settled. A Sauternes yeast strain, native to the Sauternes district of France, appears to ferment glucose more slowly than fructose and aids in controlling the rate and extent of fermentation. A variety of other wine yeasts also ferment fairly slowly.

After the must settles, adding only 0.5% of yeast culture gives a slower fermentation than adding the usual 2% would give. If fermentation is too fast it is better to cool the must than to add more sulfur dioxide.

The extent of fermentation can be roughly followed by measuring °Brix. If lab facilities are available, one should directly measure the alcohol, extract, and reducing sugar to determine exactly when to stop the fermentation. To make control of fermentation as accurate as possible, sweet wines are usually fermented in small containers, at 60° F or less. When the process is stopped, they should have 0.5–3% sugar and not over 12% alcohol.

Slowing the rate by cooling is a useful procedure in making sweet table wines via a stopped fermentation. The yeasts will be mainly in the lees, and fermentation can be further slowed by racking the must (with some aeration) off the yeast several times. Careful racking transfers only a small portion of yeast cells. They multiply, carrying on fermentation, but the multiplication reduces nutrients in the must and makes fermentation easier to stop. The addition of small doses of sulfur dioxide at each racking results in an accumulation of desirable fermentation byproducts such as glycercine. This technique is used in Italy to produce sweet, low-alcohol sparkling muscato spumante wines.

California sauterne is made by stopping fermentation before completion by means of racking and adding 250 ppm or more sulfur dioxide. Some wineries stop fermentation by centrifuging and then tight filtering. Another technique is to cool sweet wines to 35–50° F, add sulfur dioxide to give 25–35 ppm free SO_2, and remove yeasts by pad or diatomaceous earth filtration. Still another method is to add 6 grams of bentonite per gallon a few days before the fermentation will be stopped and, when residual sugar reaches the desired level, add 75 ppm of SO_2 and rack and chill the wine. A fermentation can be stopped by filtration alone but this is difficult to do since yeast cells quickly plug most filters. Filtration is most useful when wines have been racked and fermentation slowed so that the bulk of the yeasts are already removed.

When sweet wines are stored in wood they tend to lose some sugar so in that case fermentation should be stopped when the wine contains slightly more than the desired level of sugar. Sweet white table wines should be cold-stabilized. The necessary aging period for them is often 2–4 years but they should be clarified as soon as possible and bottled early. Before bottling, the winemaker should determine total and free sulfur dioxide. Total SO_2 should not be more than about 250 ppm and free SO_2 should be about 25 ppm.

EXAMPLE 12A
Producing a sweet Riesling wine

In a semi-cool climate with a long growing season Riesling grapes are harvested at 24° Brix and 0.9% acidity. In general, the procedures are those outlined in Example 10B, but there are some differences. After the juice has settled, it is innoculated with 1–2% of a Steinberg or Epernay 2 yeast culture. When fermentation starts, in a few days, the winemaker slowly reduces the temperature to about 50° F. Fermentation takes 3–4 weeks. When testing shows that the wine has almost reached the desired residual sugar level (about 6%), it is fined with 6 grams of bentonite per gallon. Several days later it is chilled to 30° F to stop fermentation, racked, and treated with 100 ppm of sulfur dioxide. The wine is kept cool and racked again a week later, or alternatively, it can be filtered through a 0.65-micron or smaller-pore filter. If the wine is not filtered, the winemaker may have to keep it cold and rack it several times to eliminate yeasts.

The wine is cold-stabilized and when it is about 6 months old it is put through a 0.5-micron filter and has 50 ppm of sulfur dioxide added before bottling. At this point the wine has about 9.5% alcohol, 0.80% acid, and 5% residual sugar. It is best consumed within 1–3 years after bottling unless it can be stored at cool temperatures, which can preserve it longer.

Botrytised Wines

To make French Sauternes the grapes are crushed in the usual way and the fermentation is allowed to proceed until the alcohol level increases to where the yeasts are killed while some sugar still remains. This is possible only in years when the proper amount of botrytis infection occurs. The result is a high alcohol content and a balanced wine made by an entirely natural process.

Botrytised grape musts ferment more slowly than normal musts probably because the higher sugar content has a retarding effect on yeast growth. Skin contact should be minimized if the grapes show mold other than botrytis. Clarifying the high-density juice is often more of a problem than with normally ripe grapes. During fermentation it is necessary to monitor volatile acidity carefully and after fermentation to protect the wines from oxidation, to adjust the amount of alcohol, acid, and sugar to the desired balance, and to inhibit laccase (as was discussed in Chapter 10).

The wine is fermented in oak and then stored in oak casks. It is racked every three months for a year, then fined at the end of the first year and again at the end of the second. After three years it is bottled.

Sweet Wines Produced by Sweetening Dry Wines

Many winemakers, especially in Germany, France, and the United States, ferment wines to dryness and then add some "sweet reserve," usually juice from very ripe grapes preserved by the addition of 200 ppm of SO_2 and refrigeration. Some wine laws require that the sweet reserve be fermented to at least 0.5% alcohol before being added, perhaps in the belief that this gives a more natural wine flavor. Grape juice concentrate is used by some California winemakers to sweeten wines, with 1–3% being added just prior to bottling. Sweet wines may also be used to sweeten dry wines.

When blending grape juice, concentrate, or sweet wine with a dry wine, one will find the blending procedures discussed in Chapter 9 useful. Sweetening agents should generally be added after the main wine has been completely clarified and stabilized. Sweetened wines can be stabilized against refermentation by means of a membrane filtration or the addition of sorbates.

Eastern U.S. and Canadian wineries often sweeten dry wines with cane or corn sugar. (Corn sugar or syrup is available in "high fructose" versions, has no detectable taste difference from cane sugar, and in large quantities usually costs less than cane sugar.) Stabilization is achieved by some combination of high SO_2 with membrane or sterile pad filtration or sorbate addition.

A unique type of sweet wine produced in the United States by several large wineries is a kosher wine made from Concord grapes; though designed to be acceptable to orthodox Jews, it is also enjoyed by many non-Jewish people. It is often made from frozen grape juice or concentrates, permitting production near major markets during the entire year (wineries producing it have been located in Brooklyn and

Chicago). Concord grape juice is generally sweetened with cane sugar or corn syrup to attain the desired alcohol level, fermented to dryness, stabilized, then sweetened with cane sugar or corn syrup to the desired sugar level. Such wines often have about 13% sugar and 13% alcohol. Some Concord wines of this type, not necessarily kosher, have been produced with 20% sugar, making them among the sweetest wines sold anywhere.

Home winemakers usually use cane sugar to sweeten wines. Corn sugar is sold in home winemaking shops for beermaking, but usually costs more than cane sugar in the small quantities many home winemakers are likely to want, and offers no offsetting advantages. Table sugar derived from sugar beets is available in some northern areas of the country, but some winemakers who prefer cane sugar claim that beet sugar gives hazes that are difficult to remove.

When sucrose (table sugar) is used to sweeten wines it tends to give them an artificial flavor for several months, until the sucrose is hydrolyzed to glucose and fructose. When one is sweetening wines in trial batches, it is sometimes preferable to add a mixture of glucose and fructose. Such a mixture can be obtained by dissolving 47.5 grams of sucrose and 1 gram of citric acid in enough water to make 100 ml, then heating the mixture nearly to boiling for half an hour. If water is added to the cooled mixture to bring its volume to exactly 100 ml, the solution will contain 0.5 grams per ml of glucose/fructose.

Some amateur winemaking literature suggests sweetening wines with other than fermentable sugars to eliminate the problem of refermentation. Among sweeteners occasionally recommended are glycerine and lactose (milk sugar). Glycerine is not very sweet, however, and at levels where it can contribute some sweetness (over 1%) it gives wines an artificial taste and feel in the mouth. Lactose also has so little sweetness compared to fructose that it is essentially worthless for sweetening wines. Possibly some people might find that using an artificial sweetener along with lactose to provide body satisfactory, but it is unlikely. It is well known that artificially sweetened soft drinks usually have less desirable flavors than those sweetened with sugar.

Preserving Sweet Table Wines

The main methods of preserving sweet wines in storage and after bottling are: (1) removal of yeasts by chilling, fining, racking, (2) removal of yeasts by means of a suitably tight—preferably sterile—filtration, (3) killing of yeasts by heating, and (4) prevention of yeast multiplication by use of sorbates. Storing wines at 34–36° F will generally prevent refermentation but is not practical for most winemakers. Home winemakers and small commercial wineries can often achieve a decent measure of yeast removal by using method (1) above, but for added safety should also include (2).

Sweet wines can be stabilized by pasteurization to about 140° F and then being held for 1–2 days at 130° F with 100 ppm of sulfur dioxide added, after which the wine can be given normal cellar treatments. An alternative is flash pasteurization to 160–185° F, followed by rapid chilling and the usual cellar treatments. Pasteurization is usually avoided for high-quality wines because of the possibility of degrading flavor. Some studies suggest that holding sweet wines in barrels at about 120° F for around two weeks can help stabilize them and smooth the flavor, but this process is impractical for most amateur winemakers.

In the author's experience, microwaves can sterilize small batches of sweet wines. Knowing that microwaves are preferentially absorbed by compounds containing oxygen, including water, alcohol, sugar, and fats, I reasoned that perhaps the fats in yeast cells would heat up more rapidly than the bulk of the wine and kill the cells with less heat treatment than pasteurization usually involves. Sweet fruit wines treated this way have not refermented in up to 4 years' time.

EXAMPLE 12B
Stabilizing a sweet wine with microwaves

Four-fifths quart of water is heated in a microwave oven until it reaches a temperature of 160° F. The time required (usually about 3 minutes) is noted; then 750 ml of a sweet wine ready for bottling is heated in the same container for the same time with two stops for stirring. If the wine temperature is not close to 160° F, the heating time should be adjusted.

The wine is removed from the oven and covered with a sheet of plastic or aluminum foil while a polyethylene funnel, a clean empty wine bottle, and a cork are heated in the oven for 30 seconds (or until the cork is softened). The wine is then poured through the funnel into the wine bottle and immediately sealed with the cork. (The jaws of the corking device should be sterilized with boiling water before use.)

The bottled wine is then placed in a bucket of cold water to cool it down to room temperature.

With practice, one can sterilize and bottle a dozen or so bottles of wine in about an hour. If heating time and temperature are minimized in this way, the quality of the wine should suffer minimum damage.

The use of sorbates, discussed elsewhere in the book, should generally be avoided because of the possibility of producing off flavors and odors. Because sorbates do not kill yeasts but only prevent them from multiplying, they sometime fail to prevent a slow fermentation in the bottle if yeasts have not been reduced enough.

Commercial winemakers generally hold sweet wines for 30 days after bottling to make sure that no yeast growth is occurring in the bottle. If less than 10 yeast cells per bottle are present at the time of bottling and the number decreases during the first month of storage, the wine is generally safe. But if the number of yeast cells remains constant or increases, then the wines must usually be held for up to 90 days to be certain that they are stable before releasing them. Amateur winemakers should check bottled sweet wines frequently to be sure they are not beginning to referment.

13
Sparkling Wines

Using care, home winemakers and small commercial wineries can produce bottle-fermented sparking wines of high quality. Bottle-fermented wines are primarily for special occasions, however. High production costs and high taxes put them out of most people's reach for everyday use. Tank-fermented and carbonated sparkling wines are less costly and are often enjoyed on informal occasions.

One way to classify sparkling wines is by the source of carbon dioxide:

(1) Wines bottled before the primary fermentation has ended. This is done in several European areas and with the rare muscato amabile wine in California.

(2) Wines that absorb carbon dioxide from a malolactic fermentation. Vinho Verde wines from Portugal and some other European wines are made this way.

(3) Wines that undergo a complete fermentation and have sugar added before a second fermentation in bottle or tank. Most sparkling wines are produced this way.

(4) Wines to which carbon dioxide gas has been added from an external source. Many inexpensive sparkling or "crackling" wines are made this way.

Among sparkling wines produced by a secondary fermentation several types are recognized in commercial trade:

(1) Wine produced the traditional way with slow bottle fermentation, long aging on the yeast, and disgorging. Identified with label statement: "fermented in this bottle."

(2) Wine produced as in (1) but transferred to a tank under pressure and filtered before being put back into bottles, a process widely used in North America. Identified with label statement: "fermented in the bottle."

(3) Wine produced without aging on the yeast and with transfer. Less expensive sparkling wines are often produced by this method. Also labeled: "fermented in the bottle."

(4) Wine produced by a secondary fermentation in a tank rather than in a bottle. Identified with label statement: "bulk process."

The producers of sparkling wines usually label them to indicate the level of sweetness. The driest are labeled *brut* or *nature,* the medium sweet *sec* or *demi-sec,* and the sweetest *doux.* Brut should contain less than 1.5% reducing sugar, sec 2–4% sugar, demi-sec up to about 5%, and doux over 6%.

Traditional Champagne Method of Making Sparkling Wines

In the traditional method of making sparkling wines, developed in the Champagne district of France (*méthode champenoise*), a dry wine with slightly less than usual alcohol is placed in a suitably strong bottle with a little sugar and a special yeast. After the yeast has fermented the sugar, built up carbon dioxide pressure in the bottle, and aged for a suitable time, it is removed, a sweet "dosage" is usually added, the wine is restoppered, and after some further aging it is consumed.

Various factors influence the quality of Champagne wine. In cool years the white grapes used (Chardonnay) do not ripen fully so red varieties (Pinot noir or some Pinot Meunier) are blended in to give a better balance of alcohol and fruitiness. Approximately ⅔ of the grapes used are red. Champagne grapes are picked at about 18.5–19° Brix. Diseased fruit is carefully removed from red grape bunches because these berries can contribute unwanted color to the white wine. The best possible grapes are used and fermentation is started as quickly as possible to minimize the need for sulfur dioxide.

The grapes are usually not crushed but are pressed several times, which reduces both color and tannins in the must. The first pressing gives the best-quality must, and for the best-quality sparkling wines only this must is used. Fermentation is usually done at 60–68° F. Chardonnay and Pinot noir (or Pinot Meunier) wines are racked several times, filtered, and then blended. A clean wine of 10.5–11% alcohol is desired. Wines for the "cuvée" or blend are aged for 8–18 months and blended just a few weeks before the secondary fermentation. Stocks from different parts of the Champagne district and from different years are used to achieve consistent quality.

In Champagne the secondary fermentation is always in bottles, preferably at 50–60° F. A calculated quantity of sugar syrup is added to the cuvée, which is then tightly filtered to clarify it and remove any lactic bacteria. A culture of alcohol-tolerant yeast is added to begin the secondary fermentation. The wine is put into bottles sealed with a hollow plastic cork topped with a crown cap. When fermentation is completed, wines are allowed to mature for up to several years. Yeast autolysis contributes a characteristic flavor during this process.

Wines are disgorged after they have spent at least 9 months in the bottle—in the case of vintage Champagnes often 3 years or more. The yeast is removed from bottle-fermented champagnes by a process called riddling. Bottles are placed neck down on special racks, at a moderate angle to begin with and later more nearly vertically. Workers dislodge the yeast by giving each bottle a sharp twist, about ⅛ turn in one direction and about ¼ turn in the other, and then dropping it back into the riddling rack. A white line painted on the bottom of the bottle helps the "turner" keep track of these motions. A good turner can handle tens of thousands of bottles per day. Wines vary in the speed with which they clear. A coarse yeast sediment may move down into the neck within a week; others take much longer. In many modern wineries this hand process has been replaced by mechanical riddling devices that tilt and vibrate bottles.

"Disgorging" is the process that eliminates yeast from bottles. Bottles are cooled to about 45° F, then placed neck down in a freezing mixture (commonly ice and calcium chloride). When the wine in the neck is frozen, the bottle is raised to a 45° angle and pointed toward a container. When the stopper is removed, pressure blows out the plug of ice and yeast. (Wine may froth out if it contains sediment or the bottle is not clean and smooth.) After the frozen plug has blown, any

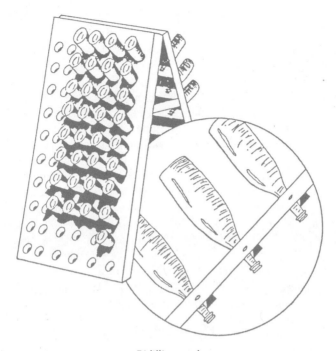

Riddling rack

yeast on the inside of the bottle neck is removed with a finger or swab, the wine is given a dosage of sweet syrup, and the bottle is filled and corked.

A "dosage" is a solution of 50–60% sugar in wine. In Champagne, where many bottles are handled and the dosage may be kept for days, about 10% brandy and up to 150 ppm of sulfur dioxide are added to retard fermentation. The dosage should be gently poured down the inside of the bottle to prevent foaming. After disgorgment, wines may be further bottle-aged before being released for sale.

American Bottle-Fermented Sparkling Wines

In California, grapes suitable for quality sparkling wines have been grown in cooler regions and are usually harvested at 19–20° Brix. Recommended varieties include Chardonnay and Pinot noir. The main grape varieties for eastern commercial sparkling wines are Catawba, Delaware, and Aurore. Ives Seedling, Fredonia, Clinton, and Concord have been used to make sparkling burgundies. Cold Duck, a sparkling wine often with a labrusca flavor, is made in the United States and Canada; it appears red, but is essentially equal parts of red and white wines.

If the white wine is too low in tannin to clear easily, a small amount (0.01–0.03%) of light-colored grape tannin can be added. Fining with bentonite is a common cellar treatment; sometimes gelatin or isinglass is used instead. The wine should be filtered and cold-stabilized before the secondary fermentation. An ion exchange may remove essential growth elements and causes a slow bottle fermentation.

Blends of different wines (called a cuvée, as in France) should first be prepared in a laboratory. The chemical and physical properties desired in a cuvée are 0.65–0.75% total acidity, pH below 3.3, alcohol of 10–11.5%, light yellow color, low volatile acidity, a clean fresh odor and flavor with no overwhelming varietal characteristic, and total sulfur dioxide below 60 ppm. Cellar treatments to achieve hot and cold stability are desirable.

Before the secondary fermentation is to begin, a yeast culture should be built up. Either dried or liquid Champagne yeast cultures are suitable. Since most yeasts are not tolerant of both sulfur dioxide and alcohol, one must usually minimize sulfur dioxide use in the still wine or reduce it by using hydrogen peroxide (see Chapter 6). One can build up a yeast culture by adding yeast to wine containing 5% sugar and shaking and aerating at 70–80° F until the sugar has been reduced to 1–2%. The usual amount of yeast starter added for the secondary fermentation is 2–3% (15–20 ml per bottle). Either the yeast/wine mixture should be agitated to suspend the yeast evenly during the transfer of wine to bottles, or measured amounts of yeast culture should be placed in each bottle before the wine is added.

In some cases yeast growth factors in the still wine will have been depleted during the original fermentation so yeast nutrients are needed for a good secondary fermentation. When that fermentation is conducted at low temperatures, cold acclimated yeasts should be used. Aerating the wine being placed in bottles will stimulate yeast growth.

During the secondary fermentation, pressure increases about 3–4 pounds per square inch (psi) per day and the process is completed in a month or so. Winemakers producing a sparkling wine for the first time or using an unfamiliar base wine should try secondary fermentation in just a few bottles and open them after 4–6 weeks to see if it has gone properly. If the wine is still sweet and the carbon dioxide too low, corrective steps should be taken before bottling the main batch.

Only champagne bottles should be used for sparkling wine fermentations. Most American champagne producers ferment in bottles capped with crown caps, cheaper and easier to apply and remove than the champage corks used on finished sparkling wines. If corks are used, they should be softened with cool water for several days. Plastic champagne stoppers can be used. Champagne corks are held in place during fermentation by a special metal clamp or wire hood.

In commercial production the pressure desired at the end of bottle fermentation is 5–6 atmospheres (75–90 psi) at 50° F. To produce each atmosphere of carbon dioxide pressure approximately 0.4% of sugar (3 grams per 750-ml bottle) is required. If the wine is not dry before the bottle fermentation, the reducing-sugar content of the still wine must be measured by analysis, and the sugar to be added lessened by the amount of reducing sugar already present. For 5–6 atmospheres the required sugar per bottle is 15–18 grams. Home winemakers are advised to aim for no more than 4 atmospheres of pressure in their sparkling wines, so 12 grams of sugar per bottle is the right amount.

The sugar is often added as a 50% solution in wine. It is sometimes commercial practice to add 1–1.5% citric acid to the sugar solution and let it stand for several weeks before use so that the sucrose has time to be hydrolyzed to glucose and fructose. (The resulting mixture is called "invert" sugar.) Alternatively, one can dissolve the sugar in water, add the citric acid, and heat it to nearly boiling for half an hour. Invert sugar, more easily fermented by yeasts than sucrose, helps ensure a good secondary fermentation.

The desired temperature for the fermentation is about 60° F. Corked bottles should be laid on their side during it but if crown caps are used bottles may be stood upright.

For disgorging, some amateur winemakers have made racks with holes that allow the necks of bottles to extend into the top of a horizontal food freezer and have been able to form an ice plug in an hour or two. Mixtures of dry ice are sometimes used, but there is a greater chance of the extreme cold causing a bottle to shatter. To avoid the hazard of explosion, winemakers should use only sound champagne bottles. After the secondary fermentation has started, personnel should handle bottles only when outfitted with proper protective equipment. Minimum safety equipment should include a protective plastic or wire mesh face shield, leather gloves, and heavy padded clothing to cover the main portions of the body. The danger is reduced if the bottles are refrigerated before handling to lessen the internal pressure.

After disgorging, home winemakers can temporarily seal bottles with the devices used to retain pressure in pop bottles until they are ready for recorking. For the final capping, plastic closures have the advantages of lower cost, cleanliness, and ease of removal. They are, however, more porous to air than cork stoppers and are seldom used on the finest sparkling wines.

EXAMPLE 13A

Producing a blanc de blanc sparkling wine

Premium white grapes, such as Chardonnay or Vidal blanc, are harvested at about 19° Brix and a total acidity of 0.75–0.85%. (If the acidity is much higher or lower, adjustments will be needed.) The grapes are carefully stemmed, crushed, treated with no more than 50 ppm of sulfur dioxide, and pressed. Only the free-run and lightly pressed juice is used for the sparkling wine. The juice is settled, given a moderately cool fermentation (below 60° F if possible), and the wine clarified. A stable, dry wine of approximately 11% alcohol and 0.65–0.75% acidity should be obtained.

A culture of a Champagne yeast is built up using added yeast nutrient, and when it is in full fermentation, 15–20 ml is added to each bottle along with 12 grams of cane sugar. The bottles are filled with wine, capped with crown caps, and shaken to mix in the yeast and dissolve the sugar. They are stored in a cool place (preferably about 60° F) and shaken every few days for about two weeks to mix the yeast with the wine. (A face shield and heavy gloves should be worn while shaking the bottles.) After a month, one bottle of the wine should be cooled in a refrigerator and opened to see if the expected amount of pressure and bubbles are present. If all is well, this bottle can be immediately recapped, and the entire batch of wine should be stored for about a year in contact with the yeast.

Before the disgorging step, bottles are riddled until all the yeast has settled in the necks. They are then chilled to 45° F and placed neck down into a mixture of calcium chloride (available from hardware stores or concrete suppliers) and ice with a temperature below −10° F so that about 1 inch of the wine in the neck is frozen into a solid plug.

The bottles are opened with a bottle opener while pointed at about a 45° angle toward a catch basin. When the plugs of ice and yeast blow out, the inside of the necks is wiped to remove any yeast, the bottles are filled with either dry or sweetened wine—depending on the final sweetness desired—and stoppered with corks, which are then held in place with wire hoods.

The wine can be consumed fairly soon but will retain its quality for several years.

Transfer-Process, Tank-Fermented, and Carbonated Sparkling Wines

The transfer process avoids the multiple and costly individual handlings of each bottle that riddling requires. Instead, the bottles of wine are chilled to reduce the carbon dioxide pressure and uncapped in a special machine (which empties them into a tank and sends their contents through a filter without loss of carbon dioxide pressure). The wine is then rebottled in clean bottles, usually with an added sweet dosage. The transfer system has two additional advantages: less wine is lost during disgorging, and wines of different ages can be blended in the tank.

Wines made by the transfer process ought to be equal in quality to those produced in the traditional process if the wine is allowed to stay in contact with the yeasts as long and if grapes of equal quality are used. Unfortunately, many commercial transfer-process wines are not adequately aged on the yeast, and frequently they are made from lower-quality grapes. Because the final wine may still contain viable yeasts, in many cases too much sulfur dioxide is used. Transfer-process wines of this type are little better than wines produced by the tank process.

The widely used tank or Charmat process is suitable only for larger commercial wineries. In this process the still wine cuvée is prepared as for bottle fermented wines. Stainless steel pressure tanks of 1000–25,000-gallon capacity are used for the secondary fermentation, at temperatures varing from 50 to 75° F. At lower temperatures the wine absorbs carbon dioxide better, and even at 50° F fermentation is over within two weeks. Although fermentation can be stopped to retain sugar, in California the wines are usually fermented to dryness and then sweetened. Following fermentation the wine is chilled to precipitate tartrates, then centrifuged and/or filtered, and bottled.

This process makes possible close control of fermentation rate and pressure, savings of labor, and short production time. But the wine must be quickly removed from the heavy yeast sediment on the bottom of the tank to prevent possible hydrogen sulfide formation, so it does not receive the lengthy yeast contact that gives bottle-fermented sparkling wines their character. More contact with air occurs in the tank process. Because it is difficult to remove all viable yeast cells from a newly fermented wine, 150–200 ppm of sulfur dioxide frequently must be added at bottling time, decreasing quality.

Tank-process wines do not retain carbon dioxide as well as bottle-fermented, perhaps because of their very short contact with yeasts, but possibly also because of their amount of contact with air. Storing tank-process wines for a year or more can somewhat increase their carbon dioxide retention. In any case, the generally lower quality of these wines seems to result from lack of lengthy yeast contact rather than from container size.

Carbonated wines are those charged with carbon dioxide artificially instead of via fermentation. One method of charging them is to chill the wine to just above its freezing point (24° F), pressure the wine-filled container with carbon dioxide and allow the gas to be absorbed as completely as possible, then bottle the wine under pressure. Carbonators of the type used for soda water can be used for wines; the pressure of the bottled wine should be about 5 atmospheres (75 psi) at 50° F.

Carbonated wines are much cheaper to make than fermented sparkling wines and can be pleasing if the wine with which one starts is of good quality. Commercial wines of this type usually have about 12% alcohol and may be acidified with citric acid and sweetened before being carbonated. Carbonated wines generally lose carbon dioxide more rapidly after being opened than do fermented wines; the latter's longer-lasting bubbles may be due to a carbon dioxide/protein complex. According to a recent patent application, deaerating a wine under slight vacuum before carbonating improves carbon dioxide retention.

Amateur winemakers carbonating wines should use plastic soda-pop bottles. Pneumatic tire valves can be cemented into plastic screw caps. With a matching quick-connect fitting on the hose from the carbon dioxide tank, wine can be carbonated (in about one minute of shaking) and the bottles disconnected from the tank without pressure loss.

Carbonation can make fairly neutral still wines more interesting. It can also be used to get a quick idea of how various still wines will taste as sparkling wines. Those that warrant the additional trouble may then be produced as bottle-fermented sparkling wines.

Alternative Methods of Producing Homemade Bottle-Fermented Wines

Several alternative methods of producing sparkling wines are used by home winemakers. The better ones are (1) leaving the yeast in the bottle, (2) trapping the yeast in a special external chamber which is then tied off, and (3) keeping the yeast inside a porous tube within the bottle. Less satisfactory methods include (4) disgorging yeast by pouring wine into clean bottles, and (5) bottling a wine that has not completely finished a primary fermentation.

Leaving the yeast in the bottle may seem a bad idea for aesthetic reasons, but it offers significant savings of effort and supplies over most other methods. By skipping the riddling and disgorgement steps, a home winemaker can produce sparkling wines almost as easily as still wines. And even bottles forgotten for 2 or 3 years will often continue to improve in contact with the yeast.

For wines that are going to be consumed away from home, as at a picnic, or entered in a wine competition, yeast removal is essential. But home winemakers should not feel required to riddle and disgorge every bottle.

EXAMPLE 13B
An easy sparkling wine with yeast remaining

A still white wine is produced as in Example 13A. The secondary fermentation is started in the same way, but the bottles are stored upright and the riddling and disgorgement steps are eliminated. When the wine is ready to serve it is carefully moved to a refrigerator or ice bucket, then opened and carefully poured.

If the wine is poured with a slow steady motion and the bottle is not tipped back to an upright position, it is possible to serve all but two or three ounces without any yeast getting into it. The final cloudy ounces can either be discarded or drunk by the host. If only two or three people are going to consume the wine it is better to put it up in 375-ml bottles so that it can be served all at once.

Home winemaking distributors sell a plastic holder that is a 1-bottle riddling rack and a special stopper with attached bladder to catch the yeast. After the yeast has been riddled down into it, the bladder is tied off and the bottle can be stored upright. If the wine is then chilled to 25–30° F, the special stopper can be removed and a champagne cork and wire hood installed. This procedure is feasible if just a few bottles require yeast removal.

Another method that offers some labor savings and is reportedly being investigated by major champagne producers, involves the use of dialysis tubing (porous cellulose) that keeps the yeast inside of a small, removable pouch. The main advantage of this method is that no special supplies are required beyond dialysis tubing (available in rolls from chemical supply houses). The major disadvantage is that if the still wine used contains viable yeast cells, the finished wine will have yeast deposits outside the dialysis pouch, so nothing is gained. A few commercial wines are marketed containing about 1.5% sugar, no sorbates, and relatively moderate amounts of sulfur dioxide. Such wines can be converted into sparkling wines via the dialysis-tube method with a minimum of fuss.

EXAMPLE 13C
An easy sparkling wine using dialysis

A roll of 1-inch-wide cellulose dialysis tubing (rated for a molecular weight cutoff of 12,000 or so) is cut into 10 in lengths, one for each bottle of sparkling wine to be produced. The tubing pieces are softened in hot water, then a tight knot is tied in one end of each. Plastic drinking straws are cut to lengths of about 5-in, and one is inserted into each piece of tubing to keep it standing straight. The dialysis tubing is filled with an active champagne yeast culture and the open end tightly knotted. The tubes are rinsed with cool water to wash off any yeast on the outside.

Champagne bottles are partly filled with a stable, dry white wine, and 12 grams of cane sugar are dissolved in each. A yeast-filled dialysis tube is inserted in each bottle, and the bottles are filled with wine and capped with crown caps. During the next several weeks, each bottle is gently inverted several times to mix the contents. After several months the secondary fermentation should be completed. (Because sugar must diffuse into the dialysis bag and alcohol and carbon dioxide diffuse out, secondary fermentation is slower than usual.)

Prior to opening, bottles are chilled for several hours. When the crown cap is removed, the dialysis bag should rise to the top, where it can be pulled out (a forceps is helpful). The wine can either be consumed at this time or the bottle can be filled with more wine (sweetened if desired) and restoppered for further storage.

In theory at least, one should be able to chill bottle-fermented wine and carefully pour it into fresh bottles without losing much carbonation or getting any yeast into the new bottles. (Siphoning is impractical because so many bubbles form in the siphoning tube that the suction is broken.) The use of Champagne yeasts that form a very sticky deposit (such as the UCD 505 strain) can increase chances of success with this method. But in the author's experience, most champagne bottles have enough roughness on the inside to cause severe foaming when sparkling wines are poured into them. Therefore the chilling and pouring method is not generally practical, although it may occasionally succeed.

Some home winemakers have rediscovered the technique of making sparkling wines first used in the Champagne district of France several hundred years ago: a Champagne yeast is used in the primary fermentation and when the Brix drops to about 0°, the wine is bottled in champagne bottles and continues into a secondary fermentation. This relatively primitive method can occasionally yield acceptable results if the winemaker takes special pains.

The first requirement is well-settled white grape must from varieties that clear readily. At the time the yeast is introduced to start the primary fermentation, about 2–4 grams of bentonite per gallon should be mixed into the wine. The bentonite will initially increase the rate of fermentation and can also raise the chances of hydrogen sulfide production, so it is imperative to keep the wine cool and to rack and aerate it if off smells develop. When the Brix drops to about 5°, the wine should be chilled to slow the fermentation further and then racked away from the bentonite and yeast lees. The racked wine, which will contain few yeast cells, will ferment fairly slowly. When the Brix falls to about 0°, if the wine is fairly clear it can be carefully racked into champagne bottles and sealed. The only thing this method has to recommend it is that if all goes well, one can produce a sparkling wine within a month. If the secondary fermentation does not proceed as expected—which is too often the case—the result will be only a hazy, insipid, sweet wine.

Some home winemakers have attempted to produce sweet sparkling wines by bottling at higher °Brix levels, but this procedure is so risky that it cannot be recommended. There is no practical way to know how the secondary fermentation is proceeding and if it goes faster than expected, dangerously excessive pressures can be built up in the bottles, some of which may explode.

Other Natural Sparkling Wines

Red sparkling wines, usually called sparkling burgundy, can be prepared by the same processes used for white sparkling wines. Reds are somewhat more difficult to referment than whites perhaps because of higher alcohol content or tannins.

Sparkling wines made via malolactic fermentation are produced in parts of Europe by overcropping vines and getting grapes with low sugar and high acid. Low sugar leads to low alcohol, and malic acid promotes the malolactic fermentation. Unfortunately, such wines are thin-bodied and it is difficult to control the malolactic fermentation to get a consistent result year after year. Also, when a malolactic fermentation takes place in the bottle, the off odors produced are not dispersed before the wine is consumed. There seems no reason for North American winemakers to attempt to pursue this method.

14

Fortified Wines

After Prohibition ended, American, chiefly Californian, production of fortified dessert wines increased. By 1950 approximately 100,000,000 gallons per year (85% of all wine sold) was dessert wine, and U.S. production of such wines exceeded Spain's and Portugal's combined. In recent years U.S. dessert wine production has decreased to about 60,000,000 gallons per year (about 15% of all wine sold). There is still, therefore, a moderately large market for dessert wines, especially those of good quality.

For many years wines with more than 14% alcohol were called fortified wines. Current U.S. regulations prohibit the word "fortified" on labels. In California there is a move to use the term "generous wines," and perhaps in time it will catch on. In this chapter the older term is used to refer to wines to which brandy has been added.

Aperitif and dessert wines are produced by the addition of alcohol (in a neutral brandy) to a base wine, stabilizing it and increasing the alcoholic content. Past U.S. practice was to raise it to 19–20%, but the trend now is toward less alcohol, because consumers seem to prefer it.

Sherry-type wines are made by adding enough alcohol to a fairly neutral white wine (called a shermat in California) to prevent bacterial growth, after which the wine is oxidized in a controlled manner.

Sherry is produced by several methods. In Spain, one traditional method employs a special surface (flor) yeast; another relies on long barrel aging under warm conditions. A method used in the United States and Canada involves heating the wine for several months in the presence of air or oxygen. Following oxidation, the sherry is brought to about 18–20% alcohol and aged.

To make Port types, alcohol is added to a partly fermented rich red wine, thereby stopping fermentation and retaining the fruity character. The wine is aged to increase complexity and bottle bouquet. White ports are made in much the same way, from white grapes.

Other types of dessert wines (e.g., Marsala or Malaga) are produced by variations on the non-flor sherry methods. Still others are made like white ports, or, in California, by the blending of different fortified wines. (Flavored fortified wines, such as vermouth, are discussed in Chapter 15.) Dessert wines owe their keeping qualities to the high level of alcohol (which by itself has a somewhat harsh and "burning" odor) and their sensory qualities to rich or pungent aromas and flavors.

Characteristics of Sherries

Sherries have a special oxidized character that, depending on production method, may be described as baked, rancio, maderized, flor, or nutty. The flavor comes from acetaldehyde, acetal, polymerized acetaldehyde resins, 2-phenylethanol, esters, wood extractives, and other components. The aroma and flavor of the grapes used is largely lost during processing.

There are three basic types of Spanish sherries. Finos are light straw-colored wines, usually with a flavor like hazelnuts, and possibly an oaky aroma and taste. Their flavor develops largely during a special type of fermentation with flor yeast. Amontillados, essentially finos that have been aged for a long time, are usually darker and often have more alcohol and extract than finos. Olorosos are amber in color and generally sweet, with 18–20% alcohol, and are produced by aging without flor growth.

U.S. sherries also have a range of flavors, depending on production methods. California regulations require dry sherries to have less than 2.5% sugar, standard sherries to have 2.5–4.0%, and cream sherries to have 4.0% or more.

Traditional Spanish Flor Method of Making Sherries

In the traditional Spanish flor process, flor yeasts grow on the surface of wine in partly filled barrels, with some air contact. The surface yeast colonies often resemble flowers in shape (*flor* is Spanish for "flower"), and produce agreeable substances which impart the characteristic fino aroma and flavor.

The leading grape variety used in flor sherries in Spain is Palomino, which has a neutral flavor and acceptable sugar and acid at maturity. Because the grapes usually have less than optimal sugar, they are often spread on the ground to dry for a few days, though this way of increasing sugar in the juice is less used today than in former years.

In Spain, yeso (plaster) is sometimes added to the grape juice during crushing and pressing, increasing the must acidity and also adding a slightly bitter and salty taste. (See Chapter 7 for more details.) Today many wineries adjust acidity by adding tartaric acid.

The alcohol fermentation uses mixed yeasts endemic to the Jerez region. One of the more important, which appears similar to the table wine yeast *Saccharomyces cerevisiae* var. *ellipsoideus,* develops in two stages. In the first stage it gives a fairly vigorous alcohol fermentation; later, when the sugar is down to 0.10–0.15%, in the presence of air it forms a film on the wine surface that oxidizes alcohol to acetaldehyde and eventually to carbon dioxide. The yeast also reduces acetic acid, which helps to prevent the wine exposed to air from turning to vinegar.

After the initial alcohol fermentation, the drier wines are fortified to 15% alcohol using a neutral brandy about 184° proof, placed in partly filled barrels with access for air, and held for several months. Wines that develop a flor film on the surface are separated from those that do not. This fermentation stage requires one to several years.

The complex flavors of Spanish sherries also result from the factional blending and aging procedure called a solera system. Spanish soleras, constructed from 132-gallon barrels (called butts) holding 110–120 gallons of wine, vary in size, but usually involve 3–5 or more tiers (or stages) of barrels. The youngest wines are placed in the top stage and once or twice a year finished wines—only ¼ barrel or less—are withdrawn from the bottom. Available space at the bottom stage is filled with wine from the next higher stage (care is taken not to disturb the yeast film), and wines from each solera stage are moved down in the same way and distributed among all the barrels in the tier below.

Soleras are housed in well-ventilated, above-ground storage buildings (bodegas). Both water and alcohol seep through barrel pores, but in the dry climate the water evaporates faster, so the alcohol content of sherries increases. When it gets much above 16%, the flor yeasts are inactivated. The aldehyde content of the sherry also increases with time. Major volatiles produced include acetaldehyde, 2,3-butylene glycol, and acetoin. Fino character disappears with long aging, and old amontillados are similar to old olorosos.

A solera system has several advantages. Introducing younger wines stimulates the flor yeast and helps its growth, and mixing wines of various ages increases wine complexity. Also, sherry taken from the solera remains fairly constant in character over a period of many years. Spanish sherries are among the most consistent wines produced in the world.

The best solera temperature for film growth is 68° F (though flavor may be better at 60° F), the best wine alcohol level is 14.5–15.5%, and the best pH is 2.8–3.4. Wine tannin should be very low (<0.01%), iron should be low, and SO_2 should be below 180 ppm. The average age of wine in the bottom stage of a solera depends on the number of stages and the rate of withdrawal and reaches an equilibrium after 10 or more years. With a 4-stage solera and 25% annual withdrawal, the average wine age levels out to about 7 years. With more stages or less withdrawal, the average age will be greater.

After withdrawal from the solera, sherries are fortified with an aged 50–50 mixture of high-proof brandy and fino to reach 18–20% alcohol and are sweetened to meet market demands. A special wine for sweetening, which adds fruitiness and softness, is made from very ripe Pedro Ximenez grapes, sun-dried for 10 days. The must, at 40° Brix or more, ferments slowly and fermentation is stopped by the addition of fortifying brandy.

To make the finished sherry, wines from many soleras are blended along with special wines: both a sweetening wine and a color wine (made by adding wine to a must that has been slowly boiled down to ⅓ or ⅕ of its original volume) are used in many blends. If necessary, acidity is adjusted by adding citric or tartaric acid. The blend is usually fined (with gelatin, isinglass, Spanish clay, bentonite, or egg whites), filtered, and cold-stabilized. Sulfur dioxide is adjusted to about 75 ppm before bottling.

Non-Spanish Flor Sherries

Some smaller U.S. wineries make a flor sherry, but only a few use the very labor- and capital-intensive solera system. Flor sherry is also produced in Australia, Canada, France, and South Africa.

Palomino grapes have been used for some drier sherry wines in California but almost any neutral-flavored, high-yielding, white grape is all right. California sherry production depends almost exclusively on grapes from the Central Valley. In many areas grapes for sherry are harvested at 22–23° Brix. In eastern North America, several of the labrusca varieties are used. Even though they are not neutral in character, the oxidation attending sherry production largely eliminates varietal flavors. Machine-harvested grapes are widely used; crushing and pressing is the same as in table wine production. Skin contact, which increases tannins, is minimized.

Shermats are adjusted to 0.60% total acidity and pH 3.4 or lower before production begins. Some California sherry producers plaster shermats at 5–6 grams of $CaSO_4$ per gallon of pressed juice. About 100 ppm of SO_2 is usually added to the grape must along with pectic enzymes. Although flor yeasts can ferment sugar, American sherry producers generally prefer normal wine yeasts which ferment faster. The flor wine yeast most widely used in California was formerly called *Saccharomyces beticus* but the name now suggested for it is *Saccharomyces fermentati*. During fermentation, the temperature is kept below 85° F. After fermentation, the wine is fined with bentonite.

Before the flor process is started, the shermat alcohol level is adjusted to about 15%, preferably when sugar is 0.2–0.5% (small amounts of sugar are needed by the film yeasts). More than 16%

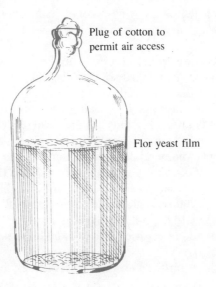

Plug of cotton to permit air access

Flor yeast film

Carboy containing developing sherry wine

alcohol inhibits the yeast and less than 14% allows bacteria to grow. Home winemakers can achieve 14.5–15.5% alcohol by using a syruped fermentation and can use a flor yeast to conduct the initial fermentation, leaving it in to form a film. Diammonium phosphate has been reported to stimulate film growth.

A portion of the yeast film can be transferred from one container to another with a clean spoon. Containers holding the sherry should be covered or stoppered so that air can enter; cotton plugs, layers of filter paper, or cheese cloth are satisfactory. The preferred temperature for flor growth is 59–68° F.

In small containers of flor wine with a high surface-to-volume ratio, fixed acids are oxidized and total acidity can decrease 0.1% or more per year.

Barrels used for sherry storage should be kept 90% full, and the developing sherry smelled and tasted every few weeks. Without air, the yeast can give volatile acidity. If volatile acidity appears, alcohol should be increased to 16% and measures taken to to get a vigorous film growth. (Actively growing flor yeasts will reduce acetic acid.) After 1–3 years in the flor stage, the wine is racked, fortified to about 18% alcohol, further aged, blended as required, and clarified. Since sherry is sensitive to oxidation it should be aged in filled, sealed barrels. Cold stabilization and filtering are common, and casein fining can be done to reduce color.

In a solera the relative humidity should be 30–40%. High humidity decreases wine alcohol, which can lead to bacterial growth, but lower humidity, which increases wine alcohol, can inhibit the film yeast growth. Some experimentation is needed to determine the best solera conditions.

Sherries withdrawn from a solera do not benefit from further bottle aging and can be consumed as soon as they are clarified.

Some flor sherry is produced commercially in North America by a submerged culture process in which wines are agitated with flor yeast. The results are similar but not identical to Spanish flor types. Acetaldehyde production of up to 400–500 ppm has been obtained in laboratories, but commercial processes are slower. Aged flavor is lacking and many of these wines are blended into baked sherries to add character. The process can produce wines ranging from the pleasantly aromatic at 200 ppm aldehyde to the intensely flavored with a heavy aftertaste at 1000 ppm aldehyde. The wine is then fortified to 17–19% alcohol and is clarified. Alcohol addition seems to intensify the flor aroma.

Despite research on the various flor processes, not all the necessary conditions for a successful flor fermentation are known, and even in commercial wineries occasional batches do not behave as expected.

EXAMPLE 14A
Producing a solera flor sherry wine

A fresh can of white grape concentrate, with cane sugar and 3 grams of diammonium phosphate added, is fermented to produce 5 gallons of wine with about 12% alcohol. When fermentation slows, small increments of sugar (totaling 1⅓ lb) are added to keep it going until alcohol reaches about 15%. When fermentation has almost stopped, the shermat is racked, fined with bentonite, and after settling, racked into gallon jugs.

A fermentation acclimates dried flor yeast to alcohol. Half the shermat is distributed in 3 gallon jugs (numbered 1, 2, 3), flor yeast is added to each, and the jug necks are plugged with cotton. The yeast grows over a period of months, and eventually a heavy white film covers the surface of the wine. After 2 more months, ⅓ of the new sherry in jug 3 is carefully siphoned (using small-diameter tubing and disturbing the flor yeast film as little as possible) into a fresh jug. The wine withdrawn from jug 3 is replaced from jug 2, which in turn is replenished from jug 1. An equal volume of shermat is added to jug 1. Every 2 months or so the process is repeated. The high surface-to-volume ratio in this small system speeds the development of flor sherry character.

If any of the jugs develops a dark color in the flor film or an "off" smell, the contents are discarded and a new jug is moved into the production sequence.

Sherry removed from the solera is fortified with alcohol to 16–19%,* sweetened with grape concentrate to taste, fined and filtered as necessary, cold-stabilized, and allowed to age for a year or more. Adding oak chips during aging generally improves character and aids clarification. If the aldehyde content is too high, small amounts of added SO_2 will reduce this smell. A final filtration before bottling improves wine clarity.

A small solera system of this type can produce about 2 gallons (10 bottles) per year. The process can be scaled up, and aging in oak barrels will generally improve quality.

(*Federal regulations, currently enforced by the Bureau of Alcohol, Tobacco, and Firearms [BATF] do not permit amateur winemakers to add spirit alcohol to wines for any purpose, even if all taxes have been paid on the added alcohol. Amateur winemakers who want to produce a stable sherry without adding alcohol might age the sherry in oak barrels in a dry room. With time, the alcohol level can exceed 16%, which is probably enough to prevent spoilage.)

Non-Flor Sherries

Some sherries are produced by long aging under warm conditions. Spanish olorosos are aged in barrels for many years after being fortified to about 18% alcohol. In the United States and Canada a few producers use long aging in barrels, sometimes outdoors (which of course further lengthens the process), to make this type of sherry.

Most competitively priced North American sherries are made by heating (sometimes called baking) the shermat in the presence of air. Baking in California is done at 120° F for 4 weeks and in New York at 140° F for 6 weeks or longer. Cooperage is stored in hot rooms, or steam or hot water coils in tanks provide heat. In the eastern states air is usually bubbled through the baking sherry to help eliminate labrusca flavor. Various methods are used to decide when baking is complete: measuring wine darkening, measuring hydroxymethylfurfural, and sensory evaluation of cooled wines.

Sugar caramelization contributes to the odor and flavor of baked wines. Winemakers can adjust shermat sweetness by stopping fermentation with fortification or by allowing it to go to dryness and adding sweet wine. After fortification, most shermats are fined and filtered before baking. For drier sherries, shermats are baked with the sugar content at which they will be marketed. Some are baked without sugar to make blending stocks. Cream sherries are usually blends of lightly baked sweet shermat and heavily baked dry shermat, but some are made by baking very sweet shermat. Sweet blends are heated after blending to "marry" the components.

Some producers use activated carbon at 0.1–7 grams per gallon to reduce off flavors or color in baked sherries. It is interesting to note that a recently patented process uses Vitamin C (ascorbic acid) as an oxidation catalyst under aerobic conditions to give, it is claimed, a superior sherry that ages quickly.

All sherries improve with aging. Young baked sherries have a strong rough taste. Oak chips (2.5–5 g/gal) are commonly used to add a mild oaky character to the less expensive ones. Better sherries are aged in 50-gallon oak barrels for up to several years. American oak is favored by both American and Spanish sherry producers because wines stored in it gain stronger aroma and drier taste.

Characteristics of Ports

"Port" originally designated sweet fortified wines from the Douro district of Portugal. Port-type wines are produced in Australia, Argentina, Canada, South Africa, the Soviet Union, and the United States. Normal ports (called wood ports in the trade) are classified as ruby (red), tawny (brown-red from age), or white. Vintage port, made in limited quantities in the best years, is aged only a year or two in oak and then for many years in bottles. These wines throw a heavy sediment (called a "crust") in the bottle and must be carefully decanted.

Ports vary widely in sugar content but average about 8–11%, with alcohol levels about 17–22%, and total acidity about 0.35–0.60%. Although California port can refer to any sweet red dessert wine with over 5.5° Brix, most California ports have sugar, alcohol, and acid levels similar to Portuguese ports. The color of North American ports varies widely because of the many different grape varieties used and variations in fermenting and aging.

Less sweet, lighter ports can be used as appetizer wines. The richer, darker, or longer-aged ports are definitely dessert wines.

Portuguese Method of Making Ports

The many kinds of microclimate in the Douro region make possible a wide range of grape varieties; about 85 are grown. Some considered most suitable for red port are Tinta Francisca, Tinta Roriz, Touriga Francesa, Bastardo, Tinta Cao, Mourisco, and Mourisco de Semente. Some leading varieties used for white port include Malvasia Fina, Donzelinho, Moscatel Galego, and Rabigato.

The traditional method called for crushing grapes by foot in shallow stone troughs (called "lagars") holding 500–3500 gallons. During fermentation regular treading assisted in color extraction. When 4–6% alcohol was formed, the must was separated from the skins and fortified with 155°-proof brandy. Modern Portuguese wineries use essentially the same methods as modern wineries elsewhere; about ⅔ of all port is made in modern facilities.

On an average, 1 ton of grapes yields 145 gallons of must, and 32 gallons of brandy is added to give a wine containing 17–19% alcohol. Higher-proof, more neutral brandies than the traditional 155°-proof are often used today. (Lower-proof brandies contribute more in aroma, but wines fortified with them require longer aging.) About 100–200 ppm of SO_2 is added at the crush to protect against spoilage bacteria and to increase the color and extract of the must. Tannin, citric acid, and tartaric acid may be added in balancing the wine. White port is made either from free-run juice or by fermentation on the skins, with the latter resulting in more flavor.

Wines are usually first racked in November although some winemakers wait until spring. The number of rackings depends on the winemaker; three the first year, and fewer in succeeding years is probably average.

Casks used for maturing wines are usually Yugoslavian oak. A "pipe," the standard cask size, holds 137 gallons. Individual small wineries deliver their newly made wines to a shipper who uses a sort of solera system to gradually blend them with older wines. Groups of pipes are called "lodge lots." Ruby ports are aged in wood for 3–8 years and tawny ports for 5–20 years. A final blend is made from the various lodge lots and sweetness is adjusted, if necessary, by adding a sweetening wine.

Red ports are fined with gelatin. They may then be cooled to within a few degrees of their freezing point, held until unstable coloring matter precipitates, cold-filtered, and treated with SO_2. Wines clarified in this way maintain clarity for years after bottling.

White ports stored in wood pick up considerable color, the excess of which is removed by a gelatin/tannin or casein fining, sometimes followed with a bentonite fining. The wines are then cold-stabilized and cold-filtered.

Most ports are exported in casks and bottled in the country of destination. (New regulations require the bottling of vintage ports in Portugual.)

American Methods of Making Ports

Many varieties of grape are used to make port wines in North America. In California, Carignane, Zinfandel, and Petite Sirah are considered satisfactory, though Tinta Madeira, Souzao, Rubired, and

Royalty are deemed better. The last three have good color, a quality often lacking in grapes grown in hot regions. White port is usually made from neutral-flavored grapes such as Thompson seedless, and is a light colored, sweet wine with no barrel aging. It may be partially decolored with activated charcoal. White ports are frequently used as bases for vermouth or other flavored wines.

Grapes for port wines are harvested in much the same way as those for red table wines except that more care is needed in handling them since they are usually very ripe (24–28° Brix). Stemming and crushing is also much the same. About 50–100 ppm of SO_2 is added to the must, shortly followed by about 1–2% of a yeast culture. Because fermentation is brief, the type of yeast used does not seem to matter much. Possibly, however, less vigorous yeasts produce a slower fermentation and hence more time for color extraction before the fortifying step. A temperature of about 60° F also prolongs fermentation time.

Punching the cap down frequently is very important, especially when it contains dried berries. A submerged cap fermenter is sometimes used, but closed tanks in which the wine is pumped over the cap several times per day are now more favored.

Ports can be made by thermal vinification, a process in which color is extracted by heating crushed grapes to about 175° F for 2–3 minutes; after cooling, the free-run juice is fermented and fortified. Dipping whole grapes in boiling water for 1 minute, to release color from the skins without heating the interior of the grapes much, is a technique that might be used for trial batches or by home winemakers. For larger batches, high-pressure steam is preferable. While heating reduces the redness of dry red wines, it usually has less effect on ports.

Within 2–6 days, the sugar content in the fermenting must is down to the proper level for fortifying—about 2% above the final desired level since fermentation does not stop immediately after fortification. A variety of presses can be used to separate the must from the grape pulp when it is time to fortify. Commercial wineries often prefer continuous screw presses for their efficiency and ease of use.

In California, brandy of about 190° proof is used to fortify port wines. Most producers now aim for 18% alcohol and 6° Brix (about 10% sugar). It is desirable to obtain these levels by fortifying at the proper time rather than by fortifying carelessly and having to blend later. The higher the initial °Brix, the lower the °Brix will be at fortification to obtain 10% sugar. Brandy is less dense than the must and tends to float on top; to stop fermentation as soon as possible, good mixing of the two is essential. Pumping over or using compressed air to force bubbles through the wine are common commercial practices. After fermentation ceases, yeasts, proteins, and gums flocculate and settle out. The wine should be racked off the lees within a week after fortification and again in about a month.

Wines that have not cleared in about 6 months should be clarified. Fining with about 2 grams of bentonite per gallon is recommended, perhaps followed by a fining with about 0.4 g/gal of gelatin. Early fining or filtering of a dessert wine will generally improve its stability. Cold stabilization at about 17–20° F for 3 weeks helps to clarify it and prevents tartrate deposits in the bottle.

Syruped fermentation, where grape concentrate is added in portions during fermentation to reach high alcohol levels, is not used commercially in the United States, but in England dessert wines are made that way and the technique may be of some interest to home winemakers. With careful syruping, alcohol levels up to about 18% are possible. The author has tasted a 20-year-old Concord wine with 13% alcohol and 12% sugar that had desirable port character, which suggests that fortification to high alcohol levels is not always necessary.

Some California winemakers make a port by blending dry wines that have good color from long skin contact with very sweet wines that have had almost no skin contact. High-quality red grape concentrate can also be used for sweetening purposes. Some inexpensive commercial ports are artificially aged by heating to 110° F for 1 week, often with the introduction of a little oxygen.

Commercial ports should be blended to a standard composition. Alcohol, sugar, acidity, tannins, and color are all important factors that may require adjustment. Close control of the time of harvest and vinification techniques can minimize the need for extensive blending of finished wines.

Ordinary port wines are stored in large tanks but the best are aged in smaller oak barrels for 1–3 years. When costs can be justified, aging goes on until the wine has reached the desired hue. The minimum amount of oak chips needed to get a threshold level of this flavor is about 2 grams per gallon.

Before the port is bottled, SO_2 should be adjusted to about 100 ppm and the wine given a polishing (sometimes a sterile) filtration.

EXAMPLE 14B
Producing a port wine

Eighty pounds of very ripe (approximately 26° Brix) Zinfandel grapes are crushed and fermented normally. About 100 ppm of SO_2 is added at the crush, and a standard yeast culture is used. Fermentation temperature is held to 60°, if possible. When the Brix drops to about 10°, enough spirit alcohol is added to raise the alcohol level to about 18%. (See the warning in Example 14A about the use of fortification by home winemakers.) After one or two days, fermentation stops, the juice is pressed from the pulp, and the wine is placed in a 5-gallon glass carboy. About a week after pressing the wine is racked, and after a few months it is fined with 0.4 grams of gelatin per gallon, cold-stabilized, filtered, treated with oak chips, and aged for 2 or more years. Aging in used wood barrels is desirable for larger batches. If the quality of the grapes was high enough, the port wine can benefit by aging for 5 years or more.

Other Fortified Wines

Other European dessert wines have been popular in the United States at one time or another. Madeira, which is made on a Portuguese island in the Atlantic by a baking process and goes through a solera, was very popular in colonial America and later in England. Malaga is produced in southern Spain from a blend of Pedro Ximenez and muscat grapes sweetened with Pedro Ximenez wine or grape concentrate and also goes through a solera. Marsala is an Italian fortified wine made by the addition of boiled-down grape concentrate and sometimes goes through a solera. In California, madeira is made by baking sweet sherry and/or angelica, malaga by sweetening sherry with grape concentrate or baking very sweet shermat, and marsala by baking a sweet fortified shermat.

A variety of other fortified wines are made in California. Angelica is basically free-run white grape juice fortified to 18–20% alcohol. To meet legal requirements, the juice is fermented to at least 0.5% alcohol before fortification. Muscatel is made much like a Portuguese white port from various muscat varieties. Fermenting on the skins for a day or brief heating is sometimes used to increase muscat flavor, though it may lower wine quality. Red muscatel is similar but uses darker colored grapes. Well-made muscatel is one of California's best dessert wines and should be of interest to some home winemakers and small wineries. California tokay is usually a blend of about equal parts of port, sherry, and angelica with sometimes a little muscatel.

15
Flavored Wines

Most of the variety of wines produced with a grape base and additional flavoring are aperitif or between-meal wines. Vermouth is the most prominent member of this category. Aperitif wines are defined by federal regulations as grape wines with 15% or more alcohol and with natural flavors. Various flavored aperitif wines with proprietary names are produced in Europe, including Dubonnet, Byrrh, and Campari. Many wines of this type contain quinine, have a sweet-bitter taste, are served "straight," with soda, or in cocktails, and are especially popular in France. When served straight they are usually iced. Retsina, flavored with resin, and May wine, flavored with woodruff, are included among aperitif wines.

In the United States the legal category of "special natural wines" having propriety names includes, among others, Thunderbird, Golden Spur, and Spanada. Sometimes called "pop" wines, they are lightly flavored with some combination of herbs, spices, fruit juices, essenses, and other natural substances. Citrus and tropical fruit flavors are popular and some wines are carbonated. They range in alcohol from about 12% up to 20%. Sales of this wine type started in the 1950s and jumped dramatically in the 1960s and 1970s, but are now declining.

Vermouths

Vermouth, legally classed as fortified wine with over 14% alcohol, is flavored with a characteristic mixture of herbs and spices that give an aromatic odor and usually a slight bitter flavor. Two broad categories of vermouths are: (1) French or dry vermouth, made from a white wine base, having about 18% alcohol and up to about 4% residual sugar, and (2) Italian or sweet vermouth, made from a white wine base usually darkened with caramel, having about 16% alcohol and up to about 16% residual sugar. The Italian type is usually heavier in flavor. Vermouths made in Europe are generally more heavily flavored than their American counterparts.

Vermouth is made commercially from fortified wine blended with an infusion made from bitter and aromatic herbs. Prepared blends of dried herbs and spices available in home winemaking supply shops might give adequate results if they were fresh, but home winemakers can do better by purchasing individual herbs and developing their own formulas. Herbs and spices used in vermouth represents parts— wood, bark, leaves, seeds, or roots—of many plants, from the tropics, the Near East, and Europe. A partial list includes:

Bitter	Aromatic	Bitter-aromatic
aloe	anise	allspice
angelica	bitter almond	bitter orange
blessed thistle	cardamom	elder
cinchona	cinnamon	elecampane
European centaury	clove	gentian
germander	corriander	juniper
lungmoss	dittany of Crete	Roman wormwood
lungwort	galingale	saffron
quassia	marjoram	sage
rhubarb	nutmeg	speedwell
	Roman camomile	sweet flag
	rosemary	wormwood
	summer savory	yarrow
	thyme	
	tonka bean	
	vanilla bean	

The quality of herbs and spices is affected by both growing and storage conditions. Dry storage at a cool temperature is best to prevent mold damage. They should be used as quickly as possible because they lose volatile components during storage and oxidation causes stale flavors. Whole plant materials generally retain volatile constituents better than powdered or granulated products.

Several methods of extraction are used. In direct extraction the flavoring materials are usually ground up, placed in a cloth bag, and suspended in wine inside a closed container for 2–4 weeks at room temperature. In California about 0.5 oz of flavoring material per gallon is used for dry vermouths and up to 1 oz for sweet. Europeans may put in 2–3 times that concentration. Grinding the flavoring material too fine may yield undesirable flavors and odors. The extraction can be done two or more times with fresh wine but later extractions generally give poorer quality. If the blend of flavoring materials is first softened in hot water it can be more readily extracted by wine, and this method may be preferable for dry vermouths. Extraction will also be speeded if the wine is heated to 140° F for one day. During extraction the wine should be stirred or agitated daily.

The goal of herb selection is a harmonious blending of bitter with fragrant and sweet herbs. Bitter components dissolve more slowly and increase as the extraction goes on. The wine should be tasted every day or two during extraction and at the first sign of excess bitterness or herbaceous character should be drawn off the herbs and filtered.

One can prepare a concentrate of the flavoring materials by using a small amount of wine and then blending this concentrate with the main batch after establishing the correct amount in small trial blends. Separate concentrates in wine (or brandy or alcohol) of each flavoring material can be used to make trial blends and develop formulas, or to balance the flavor of a batch of wine which has not quite reached the optimum.

A typical Italian sweet vermouth has a tawny-amber color, a complex, pleasing fragrance, a light muscat, sweet flavor, and a slightly bitter but agreeable aftertaste. In Italy fortified muscat wines are preferred as sweet vermouth bases; in California the base is often a fortified sweet wine such as an angelica, white port, or muscatel. Caramel is added to increase the color. Dry wine bases are sometimes used and sweetened with cane sugar (up to 20% in California) after the herb and spice mixture has done its work. Vermouth from dry wines ages more quickly and is often easier to clarify.

In Turin, Italy, the wine is usually flavored with an alcohol extract of herbs and spices that has been concentrated by partial distillation. In France direct extraction used to be common, but concentrates are now more widely used.

Citric acid and tannin are sometimes added to California vermouths to balance wine flavor, and pectic enzymes are used to clarify the wine. Pasteurization, refrigeration, and filtration are usually sufficient to stabilize a vermouth. The total sulfur dioxide content should be maintained above 100 ppm to prevent spoilage by *Lactobacillus trichodes*.

The wine base for European vermouth used to be at least 1 year old and after flavoring the vermouth was aged up to 5 years, but shorter times are now common. California vermouths are aged only about 3 months.

EXAMPLE 15A
Producing a sweet vermouth

The following formula is suggestive of the many that can be used for an Italian-style sweet vermouth.

Ingredient	Grams/gallon
Angelica	2
Bitter orange peel	2
Cinchona	6
Cinnamon	2
Clammy sage	2
Cloves	2
Coriander	4
Elder	4
Elecampage	4
Nutmeg	2
Wormwood	4

The base should ideally be a white wine with a muscat flavor, but more neutral-flavored wines are acceptable. The desired color can be obtained by adding caramel coloring or grape concentrate boiled in the open air.

The herbs should be macerated, placed in a small cloth bag, and suspended in the wine, which should then be stirred or agitated once a day for about 2 weeks. This infusion process can be repeated to flavor more wine with the same bag of herbs. If the flavor is too strong, the treated wine can be diluted with untreated base wine. If the flavor is too weak the amount of herbs can be increased or amounts of individual ingredients can be modified. Only small batches should be produced until an acceptable formula has been worked out.

After the flavoring materials are removed from the wine, it should be chilled, fined with bentonite, aged a few months, then filtered. Hazes due to pectins from flavoring materials can be removed with pectic enzymes. At least 100 ppm of sulfur dioxide should be added to the wine to prevent spoilage. Vermouths may benefit by some aging in oak or the addition of granular oak.

Since sweet vermouths typically are sweetened to about 15% sugar, pasteurization or another stabilization method should be used.

Dry vermouths, usually higher in alcohol and acidity then sweet, call for formulas with larger amounts of wormwood and bitter orange peel, but no coriander, cinnamon, or cloves. One production method used in France is to cover the herbs with fortified wine in an extraction tank and agitate the mixture frequently for about a month. The wine is drawn off and the extraction is repeated several times with fresh wine. The extracts are combined and diluted with the base wine to the desired flavor level, then refrigerated and filtered. Mistelle (grape juice preserved with alcohol) or mute (grape juice preserved with sulfur dioxide) is added to sweeten the vermouth.

California winemakers use a neutral sauterne-type wine as a base. To achieve the paleness common in American dry vermouths, decolorizing carbon can be used, but casein fining reduces wine color without removing much flavor or aroma. Sweetening with cane sugar may not mask the aroma and flavor of vermouth as much as grape concentrate does. American vermouths are bottled young (unlike French, which are often aged for 3 or more years).

EXAMPLE 15B
Producing a dry vermouth

Among the many formulas that can be used is the following:

Ingredient	Grams/gallon
Angelica	2
Bitter orange peel	6
Blessed thistle	6
Elder flowers	2
European centaury	2
Wormwood	6

The procedure in Example 15A is followed with a few changes. The base should be a dry or slightly sweet white wine of neutral flavor and little color. It can be decolorized if necessary and its flavor reduced by treating it with decolorizing charcoal (up to 5 grams per gallon). If one wants to sweeten the wine slightly by adding cane sugar, it should be pasteurized to prevent refermentation.

Other Aperitif Wines

Most proprietary European aperitif wines have 5–12% sugar and about 18% alcohol. Many are red in color and turn pink when mixed with soda and ice. Though they are made by secret proprietary formulas, a common ingredient is cinchona bark (a source of quinine) which provides bitterness. Others are vanilla, angostura, and cinnamon.

About a third of the wines produced in Greece contain added resin, and are known as retsina. After the fermentation of white table wines of good quality, 1–3% of powdered resin is mixed in for a short time and then filtered off. The resin adds an odor similar to that of turpentine, and a taste for retsina must be cultivated.

For many centuries in Europe, especially in Germany, woodruff has been added to fruity white wines, resulting in what is commonly called May wine, which has a mild aroma somewhat reminiscent of vanilla and a slightly bitter taste. The sulfur dioxide content of the wine should be low so that its odor will not interfere with the woodruff's. Fresh woodruff plants may be added to the wine directly or in an alcohol extract; other spices, including clover and ginger, can be used the same way. May wine is often served chilled as a summer drink.

Some wines with a coffee, cocoa, or other liqueur-like flavor have been produced commercially in California. The flavoring is usually added as an alcohol extract to a neutral angelica or white port base. Home winemakers can blend cocktails of sweet white wines with various liqueurs but are legally prohibited from bottling these as wines.

Fruit-Flavored and "Special Natural Wines"

As well as herbs and spices, fruit juices and other natural flavorings may be used to give a distinctive flavor to wines. Sugar, water, or caramel are all permitted in commercial wines of this type, and citric acid can be added to increase the acidity. Some wineries, especially in California, sell fruit-flavored grape wines, such as apple, apricot, and cherry, many of them noticeably sweet.

Home winemakers interested in berry or fruit wines should consider fruit-flavored grape wines, which can be produced having good body and acid balance with less effort than can many fruit and berry wines. Macerated fruit or berries should be treated with a pectic enzyme for 1–4 hours, then mixed with a white grape wine (those made from concentrates are suitable). Dried fruits can be used in some cases but lack the fragrance of fresh. After several weeks of fruit-flavor infusion, the wine should be fined, filtered, and stabilized as required. Minimum aging is needed.

A home winemaker can experiment with many flavoring materials found in the average kitchen. One of the easier to work with is the "zest" (outer layer of the skin) of orange or lemon. A small amount of zest placed in a fairly neutral white wine for several weeks will flavor it. If flavoring materials are placed in a small cloth bag they can be easily removed from the wine at the end of the infusion period. One should work with small batches until the right amount of flavoring and infusion time is found. Flavored wines can be sweetened to taste. In most cases it is easier to make a concentrated extract in alcohol (such as vodka) and blend it with the wine base to get the required flavor. Unfortunately, this cannot legally be done by home winemakers.

EXAMPLE 15C
Producing an apricot-flavored wine

The following example is a starting point; some experimentation may be needed to achieve the desired flavor.

A pound of dried apricots is put through a meat grinder and allowed to soften in a quart of water for several hours along with some pectic enzyme. A gallon of a dry, finished white wine with an alcohol content of about 14% is added and the mixture stands in a cool place for two weeks, being stirred occasionally. After the wine is racked from the apricot pulp, the pulp is squeezed through several layers of cheese cloth so that as much liquid as possible is extracted. The wine is fined with bentonite and if possible filtered through a 0.65-micron filter. If it shows a tendency to referment, one of the stabilization methods discussed in Chapter 12 should be used. The wine should be ready to drink within a few months.

———————————————

16

Nongrape Wines

Some purists and snobs contend that nothing made from ingredients other than grapes can be a true wine. The grape, the major fruit in the world, was the first to be made into wines, and much more grape wine is produced than all other types put together. U.S. regulations permit only grape wines to be labeled simply "wine" but do recognize other types.

Most winemakers would probably prefer to make grape wines if they could, but those in cold climates and in tropical or semi-tropical ones are often unable to find suitable local grapes; good imported grapes are usually very expensive and often unavailable. So people turn to other fruits and berries. In Denmark, for example, winemaking kits are based on concentrates of Danish fruits and berries designed to simulate grape wines. Although wines of this type lack grape fragrance, they are otherwise quite drinkable.

Mead, a wine made from honey that may predate grape wine, has been produced in northern Europe for centuries. It is also made in North America and might be more popular today except for the high cost of honey. Similar types of wine are made from boiled-down tree sap or sugar cane. Other bases are used, but most vegetables, grains, flowers, herbs, and leaves lack the necessary sugar and acids to make wine and are primarily used as flavoring agents in "sugar" wines. Acids and other components must usually be added to achieve acceptable results.

Fruit and Berry Wines

Wines carefully made from fully ripe fruits can often equal grape wines in quality. Such wines, usually fruity in style and semi-sweet, are sold by many wineries in North America and Europe which lie outside grape-growing districts. Often locally popular, these wines also offer something different to sophisticated wine drinkers elsewhere. Fruits that have been made into wine include apple, apricot, banana, cherry, kiwifruit, lemon, lime, mango, orange, passion fruit, peach, persimmon, plum, pineapple, prune, quince, and sloe, as well as most berries. Dried fruits sometimes used are dates, figs, and raisins. The easiest way to make a fruit or berry wine is to purchase one of the concentrates sold by home winemaking supply shops and follow directions. But such wines often lack the flavors of those made from fresh fruit, though they may be palatable if the concentrates are not too old.

Space does not permit going into detail about fruit and berry wines but some general information may be useful.

It is more difficult to extract sugar and other soluble materials from fruits and berries than it is from grapes. Equipment to chop or disintegrate the fruit, pectic enzymes, and presses that can handle the pulp are usually necessary. Large pits must be removed before fruit can be macerated. Pulpy fruits and berries can be macerated in a food processor.

Most fruits have a lower sugar content than grapes, ranging from 5% in strawberries up to 16% in some plums, and sugar must be added to the crushed fruit. Acids range from 0.35% in pears to 1.6% in tart cherries. Water is usually added to dilute excess acid. The sugar and acid contents of some fruits and berries are listed below.

Fruit or berry	% Sugar	% Acid
Apple	12	0.7 (as malic)
Plum	12	0.9 (as malic)
Cherry	9	1.0 (as malic)
Blackberry	7	1.2 (as citric)
Raspberry	7	1.6 (as citric)
Loganberry	7	2.0 (as citric)
Currant	6	2.1 (as citric)

Fruits harvested for winemaking should generally be overripe compared to those for eating or canning, and very soft. Drops are suitable provided they are not rotten or wormy. In general, fruit held in cold storage or frozen will give lower quality wines, but there are exceptions.

Cherry and pear wines tend to be protein-unstable. Strawberry and black cherry wines lose their pigments easily. Cherry, pear, and plum wines tend to brown easily. Essentially all fruit and berry wines benefit from the use of pectic enzymes and the addition of yeast nutrients.

Most fruit wines need some sugaring prior to bottling to bring them into balance and accent the flavor. It is desirable to try to duplicate the sugar/acid balance of the natural fruit and avoid over- or under-sweetness. Citric acid can be added if the acid is too low.

For fruit and berry wines, federal rules permit the addition of water to dilute the acidity down to 0.5%, and the addition of sugar, to produce not more than 14% alcohol, and also to sweeten the wine to not more than 21%. In no case is the volume of added sugar and water to exceed more than 35% of final wine volume except for loganberries, currants, and gooseberries, where the addition may be up to 60%. While home winemakers do not have to adhere to these rules, they should recognize that they were designed to ensure good quality.

Unlike grape table wines, most fruit and berry wines do not tend to improve with bottle aging, and many decline in quality within a year.

Apple wine is also referred to as cider wine. Apples contain about 12–17% sugar and 0.3–0.7% acidity. In the United States a combination of table and tart apples has generally been found best for winemaking. On the west coast a combination of Winesap and Gravenstein and in the east Delicious and Jonathan apples are considered good choices. The Northern Spy has high acidity to balance sweeter apples. Wine made from fruit stored for several months has a musty taste and is poorer in quality than that made in the fall from fresh fruit. Apple concentrate may be also used to produce wines.

Apples are sometimes stored for a few days after picking to develop aroma and then washed, crushed, and pressed. Sometimes crushed apples are allowed to stand for up to a day to develop color and flavor and make pressing easier. Pectic enzymes are usually added to the juice to improve clarity, and about 100 ppm of sulfur dioxide should be put in to retard fermentation as the enzymes work. The juice should be cooled to about 40° F and settled before fermentation.

Yeast nutrients, such as diammonium phosphate, are usually necessary for rapid fermentation. Cool fermentations (50–70° F) of apple juice are best, and Champagne yeast is preferred. The presence of SO_2 and cool temperatures help to prevent the unwanted malolactic fermentation.

European apple wines contain only about 6–7% alcohol. Commercial U.S. apple wines often have 13% and are sweetened after fermentation to about 10° Brix. The wine is usually clarified with bentonite, filtered, bottled, and pasteurized. It keeps well. In Europe and the United States some apple wines are made into sparkling wines by the Charmat process or by carbonation.

Hard cider is traditionally made by freezing cider wine and draining the juice and alcohol away from the ice crystals, a process that is unlawful for U.S. amateur winemakers because the final alcohol content can be 18–20%.

In Europe, pears of high tannin content are used for making pear wine (called perry). PVPP is used to counteract browning in pear wines.

EXAMPLE 16A
Producing a pear wine

Ingredient	Amount
Bartlett pears	2 bushels (about 60 lbs)
sugar	3 lbs (approx.)
citric acid	55 grams (approx.)
yeast nutrient	as per package directions
pectic enzyme	as per package directions
metabisulfite	3.3 grams (initially)
bentonite	20 grams
oak chips	15 grams
Champagne yeast	5 grams (1 package)

Ripe pears are picked and held at 33° F for several days until they become mushy. The fruit is grated and pressed, with rice hulls used as a press aid. The very thick juice is treated with 100 ppm of sulfur dioxide and pectic enzyme and allowed to settle for a day at a cool temperature. The juice has 14° Brix and approximately 0.25% acid as malic. Sugar to increase the Brix to 21° and citric acid to increase the total acidity to 0.55% are added. A yeast nutrient and yeast are added to the juice, and fermentation to dryness is rapid. The wine is racked, treated with 30 ppm of sulfur dioxide, stored 3 months, racked again, and treated with 30 ppm of sulfur dioxide. It is then aged for 3 months with oak chips, racked again, fined heavily with bentonite, racked, and treated with 30 ppm of sulfur dioxide. The final wine has a volume of 5 gallons and is dry. It can be sweetened to taste and pasteurized.

Cherry wines are usually best made from sour cherries since sweet ones contain too little acid. Late-picked Montmorency cherries that are immediately crushed give a good wine. Up to 10% of the pits may be broken to enhance flavor. The must should be treated with SO_2, innoculated with wine yeast, and fermented like red grape wine. Pressing is done after 2 or 3 days of skin contact, sugar is added, and the wine fermented dry. Cherry wine ferments very rapidly and may need some cooling. Fermentation at 55–60° F is recommended.

Cherry wine is often protein-unstable. Fermenting with bentonite at 2–3 grams per gallon, added when the must reaches 10–12° Brix, helps to eliminate proteins. Additional bentonite fining after fermentation may be necessary. Sugar (4–10%) should be added before bottling. Cherry wines do not keep well and develop an off flavor after about 18 months.

Black cherry or sweet cherry wine is made much the same way except that the fruit is cold-pressed before fermentation. Blending in 30% tart cherries or currants increases acidity. Pigment instability, browning, an off taste on aging, and lower fruitiness make this wine less desirable than that from tart cherries.

A wine can be made from crushed plums or sour prunes (such as Stanley) diluted with a quart of water for each pound of fruit. The fruit must be very ripe if the wine is to have an appealing color, and the pits should not be broken. Use of a pectic enzyme before fermentation greatly aids in pressing, increases the yield of juice, and hastens clearing. The must should be fermented before pressing because it is very difficult to press the unfermented fruit. The wine browns easily, so good cellar techniques and at least 30 ppm free SO_2 are needed. Plum wine is stable under normal storage conditions and improves somewhat with aging.

A very good fruit wine can be made from peaches. (One commercial winery in Michigan uses a blend of Red Haven and Harmony.) Ripe peaches are picked when they begin to drop and are crushed and pressed immediately, because browning and bitterness result from excessive skin contact time. The juice is settled and fermented cool to retain the delicate character. While peach wine can be drunk young, it ages well, developing a beautiful amber color and complex flavor after several years.

Pineapple contains about 12–15% sugar and requires the addition of sugar to give a table wine with normal alcohol content. The pineapple flavor is not stable and oxidation occurs easily, so this type of wine should be consumed young.

Orange wines tend to darken rapidly and develop a harsh, stale taste unless a high level of sulfur dioxide is maintained. The fruit should be thoroughly ripe but not overripe and the fruit must be crushed in such a way that no oil from the peel gets into the juice because the oil retards fermentation. Fresh juice should be treated with about 150 ppm of sulfur dioxide and pectic enzymes. The juice rapidly ferments to dryness. Sugar should be added to the wine to achieve about 10° Brix and enough sulfur dioxide to give 200 ppm total. The wine can be stabilized by either filtration or pasteurization. An extract of orange peel can be added before filtration to enhance the flavor. Grapefruit can be used to make a similar wine, though it is somewhat more bitter. If the juice is high in acidity, it can be heated with potassium carbonate to about 150–160° F, then filtered hot, and cooled.

Dried figs and dates can be made into wine by adding 3–4 pints of boiling water (containing 0.6% citric acid) to each pound of dry, shredded fruit. After cooling, about 150 ppm of sulfur dioxide should be added, followed by a yeast starter. Wines can be made from raisins soaked in cold water until they are plump and then crushed, pressed, and handled much like fresh grapes. The water in which they are soaked should contain about 150 ppm of sulfur dioxide. Dried apricots and peaches are not very satisfactory for winemaking because even after being softened in water they are very pulpy, gummy, and difficult to press.

A wide variety of berries have been used to make wines including blackberry, blueberry, boysenberry, cranberry, currants, dewberry, elderberry, gooseberry, huckleberry, loganberry, mulberry, raspberry, and strawberry. In the Pacific Northwest berries are usually fermented without being crushed and fermentation lasts 7–14 days. About 100 ppm of SO_2 and around ¼ of the necessary sugar are added first and the remaining sugar during fermentation.

In making wines from berries such as red currants, which have high acid, the general procedure is to add enough water to reduce the acidity to 0.8% and enough sugar to get the desired alcohol level. Pectic enzymes are often used, and sulfur dioxide should be added to the must. Most berry wines are sweetened to balance acidity and bring out flavor. Membrane filtration or pasteurization is preferred but some home and commercial winemakers add sorbate. Some berries such as blackberries tend to accumulate unstable pigments toward the end of the season and should be harvested earlier to prevent pigment deposition in the bottle.

Blueberry wine can be made by crushing blueberries in a grape crusher and adding sulfur dioxide (frozen ones can be used with no great loss of quality). Pectic enzyme should be added at 4–5 times the amount recommended for grapes. Allowing 3 days of skin contact before fermentation improves color. The berries are pressed and the juice is settled, sugared, and innoculated with yeast. Fermentation, sometimes very slow, is often helped by the addition of yeast nutrients. The process may take as long as 30 days, especially at a low temperature. Blueberry wine is one of the few fruit wines that seems to benefit from oak aging. It is stable and ages well without loss of quality or pigments.

Strawberries must be very ripe or the wine color will be disappointing, fading into pale orange in storage. Strawberries can be frozen without loss of color or flavor. After they are fermented for several days on the pulp to extract available color, the must should be pressed, with a press aid such as rice hulls. The wine can be finished in the usual fashion and lightly sweetened. It is best when young because pigment instability may be a problem.

The following example gives a recipe developed by Michael Beckmeyer; the wine won a Special Merit Award at the 1980 American Wine Society National Wine Competition.

EXAMPLE 16B
Producing a strawberry wine

Ingredient	Amount
strawberries	10 quarts (17 lbs)
water	4.25 gallons
sugar	6.5 lbs (approx.)
citric acid	120 grams (approx.)
yeast nutrient	as per package directions
pectic enzyme	as per package directions
grape tannin	2.4 grams (1 tsp.)

potassium metabisulfite	2 grams (initially)
Sparkolloid®	5 grams
potassium sorbate	5 grams
Champagne yeast	5 grams (1 package)

Very ripe strawberries are picked, any rot and the green caps are removed, and the berries are mashed by hand. The must measures 5.5° Brix and is slightly over 1.5 gallons in volume. After four gallons of water are added, the potassium metabisulfite, yeast nutrient, pectic enzyme, and grape tannin are each dissolved in 8 oz of water and mixed in. A yeast starter is prepared in a pint of crushed strawberries diluted with a pint of water to which a teaspoon of yeast nutrient is added.

Sugar to give 20° Brix and citric acid to give 0.70% (as tartaric acid) total acidity are added to the diluted must. After 5 hours, the yeast starter is added, and fermentation is active the following day. The cap is punched down daily for 3 days, then the free-run juice is withdrawn and the pulp lightly pressed. The Brix is about 11° at this time. Fermentation is completed 2–3 weeks later, and the wine is racked with the addition of 30 ppm SO_2. Further rackings are done after 2 and 4 months with 30 ppm of SO_2 added each time. After the third racking the wine is fined with Sparkolloid® and allowed to settle for a month. It is then racked, again treated with 30 ppm of SO_2, sweetened to about 5% sugar (or whatever gives the best balance to taste), stabilized with sorbate, and bottled. At bottling it has about 3° Brix, 0.73% total acidity, 11.5% alcohol, and 45 ppm free SO_2. The final volume is 5 gallons.

Mead

Honeys and the resulting meads vary greatly in flavor. Since honey contains about 82% sugar, it must be diluted with water before fermentation. During honey production bees produce an enzyme that oxidizes some glucose to gluconic acid. As a consequence, honey has a pH about as low as grape juice and resists bacterial attack. But it is not buffered as well as grape juice and when it is diluted the pH rises. To make a stable wine one must add some acid. Yeast nutrients are also required.

The example below is taken, in somewhat abbreviated form, from "Making Mead" by Roger Morse.

EXAMPLE 16C
Producing a mead

Ingredient	Amount
honey	3½ lb
water (unchlorinated)	1 gallon
diammonium phosphate	4 grams
urea	4 grams
cream of tartar	4 grams
tartaric acid	2 grams
citric acid	2 grams
Champagne yeast	5 grams (1 package)

A strongly flavored honey, such as from goldenrod, works well in making mead. (Some other authors recommend mildly flavored honeys.) The honey/water mixture should be boiled for 10–20 minutes to coagulate proteins. When it is nearly cooled, the other ingredients are dissolved, and when the mixture is at room temperature, the yeast is added. The fermentation should be over in 3 weeks; the wine still contains sugar at that point. It is racked and about 30 ppm SO_2 is added. After 3 and 6 months it is racked again, with 30 ppm SO_2 added each time. The mead can then be bottled. For improved clarity, it should be fined with bentonite or another fining agent.

Spiced meads, called metheglin, were once popular for adding flavor to mild meads and perhaps for hiding off flavors. Various herbs and spices can be used to flavor mead if they are soaked in it during or after fermentation for up to about 24 hours (longer soaking may result in bitterness). Generally two or three herbs or spices make for a more interesting flavor than just one. Combinations mentioned by Roger Morse include hyssop and sweet woodruff; rosemary and thyme; cardamom seed, lemon mint, and sweet woodruff; camomile, rose hips, Asiatic ginger, and sweet basil; and cinnamon, cloves, and nutmeg.

Fruit juices have been added to mead for centuries, apple being the most popular, with tart varieties preferred. Morse recommends adding 15–20% apple juice after the honey/water mixture has been boiled. Other juices recommended include pear, plum, peach, and raspberry.

To make a sparkling mead one should use only about 2¾ lb of honey per gallon of water so that the wine will be dry. Cane sugar is used for the second fermentation so that no added protein from honey will cloud the sparkling wine. (See Chapter 13 for details of sparkling wine production.)

Birch sap, sycamore sap, and molasses have been used in making wines somewhat similar to mead. Homer Hardwick's book, *Wine-making at Home,* gives some recipes.

Other Nongrape Wines

Home winemakers have used a great variety of plants. Vegetables from which wine has been made include beet, cantaloupe, celery, muskmelon, onion, parsnip, potato, rhubarb, tomato, and turnip. Grains include barley, caraway, malt, oat, rice, sorghum, and wheat. Flowers include carnation, clary, clover blossom, coltsfoot, cowslip, daisy, dandelion, elder blossom, marigold, primrose, rose, verbena, and violet. Herbs include comfrey, fennel, ginger, mint, rue, and sage. And leaves include balm, grape, sarsaparilla, spruce, and walnut. Some books give recipes for making wines from such ingredients that are actually sugar wines; the ingredient for which the wine is named is just a flavoring. More sophisticated recipes call for the addition of fruit acids, yeast nutrients, tannin, and sulfur dioxide. Wines of this type are seldom as balanced or interesting as grape, fruit, or berry wines but can be reasonably palatable. Some recipes call for the addition of raisins or other dried fruits to increase balance, body, and flavor. These wines may require fining or filtering for clarity.

EXAMPLE 16D
Producing a dandelion wine

Ingredient	Amount
dandelion blossoms	6 cups
raisins (white)	1 lb, chopped
water (boiling)	1 gallon
sugar	2 lbs
citric acid	18 grams
yeast nutrient	as per package directions
grape tannin	0.6 grams (¼ tsp.)
metabisulfite	0.5 grams
Champagne yeast	5 grams (1 package)

Only the dandelion petals are used, not the stems. Boiling water is poured over the flower petals and the chopped raisins. Sugar and acid are added and the mixture stirred to dissolve them. When the mixture has cooled to under 100° F, the remaining ingredients (except the yeast) go in and are also dissolved during stirring. After several hours the yeast is added. After 3 days of fermentation, the flower petals and raisins are strained out, and the wine is placed in a gallon jug with a fermentation lock and racked every 3 months with 30 ppm of SO_2 added each time. When the wine is clear, in 6–12 months, it is bottled.

Winemakers should recognize that any recipe is only a starting point on the road to satisfactory results. Most wines can be produced as dry wines but some benefit from sweetening. Sugar adds body and tends to mask other flavors. Brown sugar also adds flavor but makes the wine more difficult to clear and often necessitates longer aging.

The main requirements for any wine are (1) sufficient alcohol to help preserve it and give it a winelike character, (2) enough acidity to provide essential wine tartness and prevent spoilage, (3) nutrients for proper yeast growth, (4) pleasing fragrance and flavor components.

To develop a good nongrape wine recipe: (1) start with a general recipe for the wine type; (2) check sugar and acid before starting and make careful adjustments based on actual measurements; (3) add missing components needed for a good fermentation and a balanced wine; (4) clarify and stabilize the wine as required; (5) keep detailed records at each step; and (6) critically evaluate the finished wine.

17

Sensory Evaluation of Wine

Sensory evaluation of wines involves careful looking, smelling, tasting, and feeling—using all the senses that wines affect. Attention to detail and serious practice are needed before a person can become adept at this important technique.

To guide quality control in winemaking, sensory evaluation should be used from the start of fermentation until the last wines from a bottling are opened. Laboratory tests, which usually focus on one aspect of wine quality at a time, are more useful than sensory evaluation for certain tasks (e.g., measuring the sugar content of musts). But for subtle distinctions among colors and odors and for the overall assessment of wine quality where many factors must be considered at the same time, sensory evaluation is indispensable.

This chapter considers various sensory evaluation techniques—though there are others that are more complex, the ones presented here are adequate for most purposes—and suggests some exercises that can be used to sharpen one's skills in recognizing subtle distinctions.

Equipment and Conditions Needed for Sensory Evaluation

Serious sensory evaluations should be conducted in a proper setting with proper equipment. Wine cellars are seldom appropriate because they are too dark and often have smells which can confuse the evaluator. A good setting has proper lighting, a table with a white covering for each wine evaluator, and a minimum of noise and other distractions. Equipment needed includes several wine glasses (approximately 8 oz in size), water to rinse the mouth between wine samples, a receptacle in which to spit wines after they have been tasted, a suitable light source (such as a candle) to help evaluate the clarity of wines, and writing materials. Evaluators should keep detailed notes.

It is an advantage to have an assistant who can pour wines into coded glasses and take care of other chores. This helps to keep the attention of the evaluators focused on their task and reduces the possibility of bias.

The timing of an evaluation session is important. Most people are at their best just prior to a midday or evening meal. Sessions should generally deal with no more than a dozen samples and last no more than an hour. When important decisions must be made, several sessions should be scheduled on different days and the wines tasted more than once in different orders. Extra bottles of each wine should be available in case evaluators find a particular wine much better or worse than they expected and wish to try another bottle.

Evaluating Color and Clarity

Three aspects of color are important in evaluating wines: hue (the red, yellow, or other pure color), strength of color (paleness or darkness), and purity of color (absence or presence of orange, brown, or other tints in addition to the main color).

To gain better appreciation for your personal ability to detect changes in color strengths, fill three wine glasses half full of water and add 2 drops of red food coloring to the first, 4 drops to the second, and 8 to the third. Swirl the glasses to mix the colors evenly, then careful compare their strengths. You will probably notice that the ratio of the strengths of the colors does not double as you might expect between glasses 1 and 2 and between 2 and 3. Our senses are often nonlinear. You can try adding more color to glasses 2 and 3 until you decide that the color in each is twice as dark as the glass before it.

Detecting color purity is another important skill. Those who are color blind may find this difficult or impossible, but it is worthwhile for each person to find his or her limits. To come to an understanding of your abilities in this area, fill three wine glasses half full of water and add 5 drops of red food coloring to each. Then add 1 drop of yellow food coloring to the second glass and 1 drop of blue diluted 1:3 with water to the third (pure blue coloring is too strong). Swirl the glasses to mix the colors. Tilt the glasses and look through them at a white surface. If you can tell the difference, you can proceed to a more difficult task.

Have an assistant select one of the glasses and, without telling you which one it is, ask you to name it. Although our eyes are very sensitive to small differences in comparative color intensity and purity, we find it much harder to make decisions based on memory alone. It is much easier to make color judgments when you have several wine samples side by side.

After some practice with these artificial mixtures you can work with actual wines, doctoring them with small amounts of food colors or larger amounts of water, with the aim of learning how to detect differences and describe them in terms of hue, strength, and purity.

Young white wines are usually a very pale yellow color, sometimes with a hint of green. As white wines age their color darkens and gradually turns toward brown. White wines are darker when excessively exposed to air, when made from very ripe grapes or grapes affected by botrytis, or when aged in oak. Dessert wines such as sherries often show dark colors. When one knows the type and history of a wine, one can make decisions on whether the color is normal or not.

Young red wines are often close to pure red but may have an orange or purple tint depending on the grape variety. With aging, the color tends to lighten and turn brown. Careful observation of wine color can give useful clues about age and condition.

Sound, well-made wines are generally brilliant in appearance. Wines that look cloudy, hazy, or dull are usually spoiled with a bacterial infection or contain excess pectins, proteins, or metal salts. Wines between brilliant and dull are often described as clear. It is important to be able to judge clarity so that one can decide if a wine requires cellar treatments. To understand the basic distinctions here, do the following: Fill 3 wine glasses with 4, 3, and 5 oz of water respectively. Add one drop of milk to the third glass and swirl to mix it in. Then take one oz from the third glass and mix it into the second glass. You will now have samples representing brilliant, clear, and dull wines. You can most easily detect the clarity differences of these samples by holding each glass up and looking through the liquid at a small light source such as a candle or a distant light bulb. The third glass will give a hazy view of the light. The second will be clearer, but the outline of the light source is not as sharply defined as with the glass of plain water.

Following this exercise, one can use samples of a white or rose wine to get more practice judging clarity. If the wine has significant color, more milk may be needed to make the differences obvious.

Some wines, especially whites that are chilled before serving, may contain small crystals of potassium bitartrate in the bottom of the bottle or on the underside of the cork. This natural substance does not affect wine quality. Some older red wines may contain a precipitate or sludge of oxidized pigments and tannins on the side of the bottle which has been lowest in the bin. This material has a bitter taste and the wine should be decanted to remove it. This natural sediment is not a defect, however, and will not affect clarity unless the wine is unduly shaken up.

Other Aspects of Wine Appearance

Young wines, especially whites, sometimes show small bubbles in the glass. These usually indicate dissolved carbon dioxide left from fermentation and are a sign of youth. In slightly hazy red wines, bubbles may signal a bacterial fermentation in the bottle.

When wine is swirled in a glass, one can observe little rivulets rising and falling on the walls. This action is caused by evaporating alcohol that raises the local surface tension, allowing some of the wine to rise until its weight pulls it down. You can observe the variation in the little rivulets—sometimes called "tears" or "legs"—on the walls of a wine glass by putting 3 oz of table wine (with about 12% alcohol) and 1 oz of water in one glass, 4 oz of the same wine in a second glass, and 3½ oz of wine and ½ oz of brandy or other distilled beverage in a third glass. If you then swirl the glasses and allow the wines to settle for a few seconds you should see differences in the number of these tears. The higher the alcohol content of wines, the more tears.

Evaluating Fragrance

Flavor is the combined sensation of taste and odor perceived in the mouth. Flavor is greatly influenced by odor; without odor an apple and an onion would have similar effects in the mouth. Wine odor can be divided into aroma, derived from the grapes, and bouquet, derived from changes during winemaking and aging.

In addition to the sense of smell, the nose has a "general chemical sense"; this is what is at work when we experience a sharp burning or itching feeling in our noses when we smell household ammonia or bleach. A few components in wines (e.g., alcohol, sulfur dioxide), when present in excess, can activate the chemical sense. When the odor of alcohol is too strong it seems hot or sharp.

To experience the smell of alcohol, pour a little vodka into a wine glass and sniff it. Warming the vodka will increase the sharp chemical sensation. To learn the effect of high alcohol in a wine add ½ oz of 100-proof vodka to 4 oz of a table wine having 12% alcohol, raising the alcohol to about 16%. Smell and taste and wine and see if you don't find the result less pleasant than the original wine.

A variety of desirable odors are found in wines, foremost among them odors characteristic of the variety of grapes used.

Some eastern North American wines based on labrusca varieties (e.g., Concord, Niagara) possess a distinctive odor partly due to the ester, methyl anthranilate. This chemical is used in many artificially grape-flavored food products, such as candies and chewing gum. Smelling and tasting Welch's Grape Juice, which is made from Concord grapes, will give you a good idea of the labrusca character. The main disadvantage of this particular odor and flavor is that they are overpowering and hide other attributes of the wines in which they are found. Eastern American winemakers have learned in recent years how to minimize this odor and flavor.

In the southeastern United States a number of wines made from the native muscadine grapes have a distinctive aroma caused in large part by 2-phenylethanol, the same chemical that gives roses their distinctive odor.

Most wines produced elsewhere in the world, including California, are made from vinifera varieties and possess such a wide variety of fragrances that it is impossible to describe them all. Most fragrances are combinations of very many minor components.

One of the volatile chemicals that make wines from muscat grapes distinctive is linalool, which has a character like perfume. Asti Spumanti is made from muscat grapes in Italy, and other muscat wines are produced in the United States and Germany. This grape is believed to be among the most ancient of the vinifera varieties, and the muscat character is present to a small extent in the Riesling grape and more noticeably in the Muller-Thurgau hybrid (a cross between Riesling and Silvaner or perhaps two Rieslings), now the predominant German hybrid variety.

Most of the subtle fragrances of vinifera wines are also found to varying degrees in other plants, and it is sometimes useful to characterize wine odors in these terms. Some French Chardonnay wines have a smell and flavor akin to apples. The smell of some botrytized German Riesling wines is similar to apricots. One producer of Cabernet Sauvignon in California got a pear character in his wine, and the Cabernet Sauvignon odor occasionally resembles that of bell peppers (the same chemical has been found in both). Certain aldehydes found in leaves are also found in some hybrid grapes from eastern North America and give a grassy taste to wines.

Off Odors

Undesirable odors occasionally occur in wines. Two of these, acetaldehyde and excessive sulfur dioxide, are more common in white wines than in reds because red pigments combine with them. Since acetaldehyde and sulfur dioxide also combine with each other, these particular odor defects do not coexist.

Acetaldehyde, which results from air oxidation of ethanol, usually indicates that the wine has been abused or is too old, and often goes together with a somewhat brownish color. In certain wines such as fino sherries, it is a natural and desired component, which contributes to the "nutty" flavor. A few people seem not to mind moderate oxidation in table wines, and some seem to expect it in French white Rhones and Hungarian Tokays.

One way to learn to identify an oxidized odor is to compare the smell of a fino sherry with a white table wine. An oxidized smell can also be obtained by placing an ounce or two of white wine in a stoppered bottle in a warm place for several weeks. By mixing small amounts of sherry or oxidized wine with a fresh white wine, one can learn to detect beginning oxidation.

People vary in their sensitivity to sulfur dioxide. Present in excess, it can activate the chemical sense, give a tingling sensation in the nose, and even cause sneezing. The smell is reminiscent of lighted stick matches. It can also affect wine flavor and gives a musty character. It is most noticeable in some white wines, especially sweet ones, where it may have been added to retard further fermentation. Certain German wines so regularly contain excessive sulfur dioxide that some people consider the odor part of the "German character." Fortunately, sulfur dioxide is quite volatile and swirling the wine in a glass diminishes the odor.

To experience the smell of sulfur dioxide, add 0.2 gram of potassium metabisulfite or sodium bisulfite to a quart of water containing 5 grams of tartaric or citric acid. Since there is nothing in the water to combine with the sulfur dioxide, all of the SO_2 (approximately 120 ppm) will be free. This amount will be greater than that in any properly made wine. By diluting the solution with various amounts of water you can get some idea of your threshold for free sulfur dioxide.

Among the off smells found in red wines, high volatile acidity is probably the most common, followed by mercaptans and sulfides. Some wines occasionally have off odors caused by bacterial infections.

All wines contain small amounts of acetic acid and ethyl acetate, which are volatile and contribute to odor. In small quantities (less than 0.05%), volatile acidity does not detract from quality, but at higher levels it signals that the wine is turning to vinegar. You can, in fact, learn to detect the odor of acetic acid by smelling ordinary distilled white vinegar. In wines with excessive volatile acidity the acetic acid is accompanied by ethyl acetate. Together they have a more complex odor that you can experience by smelling wine vinegar. Try adding 10 or 20 drops of wine vinegar to 1 oz of sound red wine and see if you can detect the volatile acidity in the mixture. Good wine vinegars have modest amounts of ethyl acetate so they usually have a more pleasant odor than wines spoiled by wild acetic bacteria.

A few wines, especially reds, have a skunky odor caused by certain sulfur compounds. Yeasts can reduce various forms of sulfur to hydrogen sulfide (rotten-egg smell). In time hydrogen sulfide in wine is converted to mercaptans (skunk-like smell) and disulfides (sewage smell). Even traces of these odors can ruin a wine.

A variety of other off smells are caused by bacterial infections. Lactic bacteria can give wines a sauerkraut odor and, if sorbic acid is present, a geranium odor. One bacteria-derived odor has been described as "mousy," presumably smelling like an animal cage; others are similar to heated or rancid butter. Some wines have an earthy odor, perceived in the mouth, perhaps due to a bacterial infection. Wines made from grapes attacked by mildew can have a moldy or mildew odor. Certain molds attack corks, resulting in wine smelling in a way that has been referred to as corky or corked.

Evaluating Taste

There are three primary tastes in wines: sour, sweet, and bitter. (Few wines have a salty taste). A mild sourness due to natural fruit acids, is one of wine's most distinctive qualities, setting it apart from most other beverages. This slight sourness is one quality that makes wines go well with meals, for it moderates the greasiness of many main courses. Although the tongue can detect sourness all over its upper surface, the area most sensitive to this taste is along the sides.

If you dissolve ¼ teaspoonful of tartaric acid in 6 oz of water, the resulting solution will have the same total acidity as most wines but will taste more sour because other wine constituents, which partly mask sourness, are missing. Adding a teaspoonful of sugar to this acid solution makes the sour taste much less noticeable to most people.

Appreciation of sweetness is universal among children and is probably instinctive. It is only later in life that we develop an appreciation for foods and beverages with noticeable sourness, bitterness, or saltiness. Sweetness comes from sugar and also from alcohol, glycerine, and other wine components. Sweet wines contain between about 2 and 20% sugar. Most people can taste sugar in a wine when it is above 0.5–1.0%. The tip of the tongue is most sensitive to sweetness. A survey done at the California State Fair in 1955 suggest that about half the American population prefers noticeably sweet wines over dry ones. Sweetness in wines is neither good or bad by itself but can hide other flavors if present at too high a level. Many commercial U.S. winemakers have discovered that wines with sweetness near threshold levels are acceptable to more consumers than the same wines in a totally dry style.

A 3% sugar solution (a level teaspoonful of table sugar dissolved in

Sensory evaluation

4 oz of water) contains the amount of sugar that marks the approximate dividing line between the sweetness acceptable in table wines and the sweetness that places wines in the dessert category. You can investigate your sugar threshold in wines by setting up four glasses with 3 oz of mild-flavored dry white wine in each. Leave the wine in the first glass unsweetened, add ⅛ tsp of sugar to the second glass, ¼ tsp to the third, and ⅜ tsp to the fourth. This will give wines with 0, 0.5, 1, and 1.5% sugar. Have an assistant take these glasses away and make a code mark on each of them. Taste all the samples several times and record your impressions of their sweetness. If, for example, you can tell the unsweetened wine from the 0.5% sugar sample at least half the time, then your sweetness threshold for that type of wine is about 0.5%. You should also try to tell the 1% sugar sample from the 1.5% sugar sample in order to determine your difference threshold for sweetness, which may be different from your initial threshold. By repeating this process with several types of wines, both white and red, you will learn your ability to detect sweetness and your preference for sweetness in wines. If you find that the differences are too easy to detect, you can prepare intermediate samples. To make certain your starting wine is dry, test it with a Clinistix® test strip (or other urine sugar test product) available from a drug store. A negative or light color reaction is acceptable.

Bitterness in wines comes from tannins extracted from grape skins, seeds, and stems. Most white wines lack bitterness, but some young reds and certain dessert or aperitif wines are noticeably bitter. This taste seems to prevent the appetite from being satiated so slightly bitter wines are appropriate before or with meals.

Bitterness is detected more toward the back of the tongue than other tastes, so you will probably have to swallow a wine really to notice it. It tends to linger longer than the other tastes and contributes to the "aftertaste" or "finish" of wines.

To investigate the bitter taste, obtain some grape tannin from a shop selling home winemaking supplies. Dissolve ¼ tsp of this powder in 3 oz of water to give a solution having approximately the bitterness of a young red wine.

To learn to distinguish the three tastes found in wines, mix various amounts of the acid, sugar, and tannin solutions just prepared. (They tend to spoil so use them within a day or two.) You will find that both sourness and sweetness tend to mask bitterness; a fact that probably explains why some people add sugar or lemon to tea and why some young red wines taste more mellow when they are slightly sweet.

Astringency is not a taste but a "dry" or "puckery" feeling in the mouth. Since it is often found in red wines along with bitterness, it is worth learning to identify. To experience the feeling of astringency, dissolve ¼ tsp of alum (available in groceries) in 3 oz of water and taste the solution. Your experience will not be that of pure astringency, since alum also has a salty taste, but you will get the idea. Compare the alum and the tannin solutions until you have a clear idea of the difference between astringency and bitterness.

The "body" of wines is an often misunderstood concept; some people confuse intensity of flavor with body. One rough definition is "the lack of watery character," but more accurately, body is a feeling of thickness in the mouth due to viscosity. All dissolved substances in wine contribute to body but in dry wines alcohol contributes most. Sugars in sweet wines also contribute significantly.

"Balance," a term frequently used in describing wines, is the name of a somewhat subjective concept referring to a pleasing ratio of acidity, sweetness, and bitterness in a wine. One way to investigate balance is to start with a wine that you like and add small amounts of tartaric acid until the result is less pleasant than the original. Repeat this process adding sugar and again adding tannin, using fresh wine samples each time. Once you have a good idea of what an out-of-balance wine tastes like, you can go further.

You can investigate the effects of one taste component on another by starting with a white wine that is too sour (either naturally or because you added some acid) and add sugar until the overall taste seems improved. You can go the other way and start with a wine that is too sweet and add acid. With a red wine you can start with excessive bitterness (either natural or because you added tannin) and add sugar until the overall taste seems improved. If this causes the wine to taste more mellow but out of balance, you can add some acid to see if this helps.

It is no coincidence that some of the most popular table wines sold in the United States are semi-sweet; many consumers prefer this taste. A lot of semi-sweet wines also have more than normal acidity and, in red wines, more than normal tannins. Some commercial winemakers have balanced the taste of such wines to the point where wine writers describe them as too bland. This suggests that there are balance points where sweetness, acidity, and bitterness mutually mask each other and some consumers may prefer wines a bit out of balance.

Balance can also refer to harmony among all the sensory impressions a wine makes. For example, a Riesling wine with a golden color and definite sweetness usually comes from ripe grapes, often infected with botrytis. If a deep-colored, sweet Riesling wine has a weak smell of underripe grapes, it will appear out of balance. One needs to consider all sensory perceptions to tell whether wines are properly made.

Wood Character

Many premium red and dessert wines and a few whites are aged in oak. Oak aging increases the complexity of fragrance and flavor and also adds some tannins and coloring matter. A few wines have an excessively oaky character, which detracts from their natural qualities.

One of the desirable substances that wines extract from oak is vanillin. The faint smell of vanilla is a sign to the experienced wine taster that a wine has been aged in oak.

To experience the taste of wood in wine, obtain some oak chips from a home winemaking supply shop, place a teaspoonful in 4 oz of wine in a small container, allow this to sit for a day, then smell and taste. The odor and flavor of wood should be quite noticeable. A series of experiments can be carried out to find the amount of oak flavor that seems best for various wines.

To learn what a vanilla smell is like in a wine, dissolve a drop of vanilla extract in a tablespoonful of water and add a few drops of this mixture to 4 ounces of red wine. It may be necessary to vary the amount of vanilla extract depending on how concentrated it is. When the vanilla smell is just barely noticeable, most people find the result pleasing.

Overall Wine Quality

Overall wine quality is rarely defined, though there is a concensus among wine experts that there is such a thing. People perceive quality in different ways, however, depending on their past experiences. One definition of wine quality is "whatever it is that causes one to find pleasure in a wine." Complexity contributes if it is not excessive. A wine of great quality will be memorable, and this is a practical test. It is instructive, therefore, when sampling several wines at the same time, to pick out those that give the greatest pleasure, then set them aside and try to remember what it was about each that made a favorable impression. Doing this exercise repeatedly can help individuals define their personal standard of wine quality.

18
Wine Competitions

The wide variety of commercial and amateur wine competitions held throughout the country have a certain influence in winemaking practices.

Each November since the mid-1970s the American Wine Society has held a national amateur competition that includes Canadian wines. The Home Wine and Beer Trade Association also holds a national amateur competition. In the commercial field there is currently no national and only one large regional competition, the one held each autumn by Wineries Unlimited for U.S. and Canadian wines east of the Rocky Mountains.

Commercial wine competitions are held in several states. Michigan and New York each have one for in-state wines in conjunction with their respective state fairs. (California's state fair pioneered this type of event but has discontinued it.) Several states have amateur competitions and in some cases (e.g., Indiana and Wisconsin) wines from other states can be entered so these are actually regional.

Some of the largest commercial wine competitions are at the county level; Los Angeles County, Orange County, and others in California hold them. Some county fairs feature amateur competitions, including Sonoma County in California and Cuyahoga County in Ohio. There are many other local amateur competitions around the country, and their number is growing.

Typical Competition Rules

Each wine competition sets its own rules, generally designed to avoid unfairness and to limit entries to a given geographical area.

Commercial contest rules often limit entries to grape wines. In Michigan and New York wines are currently required to be made with a minimum of 75% in-state grapes. Most competitions require 2 or more bottles of wine to cover breakage, spoilage, or a final judging round. Most do not limit entries by vintage year or attempt to restrain previous winners from reentering. An entry fee is frequently charged. Winners usually receive a ribbon, medal, or certificate.

Amateur rules seldom discriminate against fruit wines or wines made from out-of-state ingredients, but previous winners may be barred from entering the same wines a second time. Most competitions here also require 2 bottles, and some charge a nominal entry fee. Awards, often physically impressive, include dishes, plaques, and trophies. One Maryland competition has as a rotating prize a hand-crafted sterling silver garland.

EXAMPLE 18A

Rules for entrants in the AWS Amateur Wine Competition (abstracted)

(1) All wines must be produced from grapes, grape concentrate, fruit, fruit juice, flowers, or honey.

(2) An entry fee is charged for all wines.

(3) Two 25-oz bottles of each entered wine must be delivered to the competition site.

(4) Entrants have a code number assigned to each wine which they should record and keep track of.

(5) Entrants and other interested persons act as judges in the first blind judging round. Finalists submit a second bottle for judging by a select panel (on which no entrants sit).

(6) It is the responsibility of entrants to select the proper category for their wines. Depending on the number of entries, categories will be combined or subdivided to provide a reasonable number of entries in each for judging purposes. Up to 3 awards will be given in each category if the wines merit them.

(7) No entrant can be involved in any way with commercial wine-making since this is an amateur contest.

(8) Decisions rendered by the rules committee are final.

Judging Panels

Judging panels in both commercial and amateur wine competitions in the United States are often the subject of controversy, partly because competitions are fairly new in many parts of the country and judge training is just getting started. Moreover, outside of California, judges face many types of wines besides the usual vinifera varieties. In England and Canada, where amateur competitions have been held for many years, organizations have been built up to train judges and to slowly elevate them to the top jobs. In Australia, the same sort of apprenticeship system exists for commercial competitions.

In most U.S. wine competitions the organizer starts by inviting people that he or she knows are familiar with wines. Over a period of time the better qualified and more interested judges tend to repeat, and eventually most competitions build up a respectable pool of judges.

The number of judges varies. Most commercial panels have 3–5 members. Larger numbers tend to give more reliable results, but bigger panels increase expenses and tend to make the competition unwieldy. If the competition has a preliminary judging to select the better wines and a final judging to pick the award-winning wines, all judges from all panels often participate in the final round.

EXAMPLE 18B
Judging panels at the 1980 New York State Fair Wine Competition

Judging panels have 5 members. Each judge, working independently, rates all wines in a given category and determines the awards he or she wishes to give. When the judges have finished with a category they gather, and a panel coordinator summarizes individual opinions. The judges then discuss their evaluations, and if some wish to retaste wines they do so. When this discussion ends, the panel coordinator determines the highest award that a minimum of 3 judges can agree

upon. Wines winning gold medals are retasted by all judges to determine the best-of-show award in three general categories: native
American varieties, hybrid varieties, and vinifera varieties.

Judging Procedures

In some judgings, wines are ranked and awards given to those a
consensus of the judges decrees the best. In other judgings wines are
scored on various attributes, and winners selected from the highest
average scores. In both types, it is possible to have an elimination
round in which wines ranked or scored lower are set aside and the
best wines are rejudged.

Experience has shown that most wine judges can differentiate 5–7
levels of quality. If a 5-point scoring system were used, many judges
would hesitate to award a 1 or 5, and the effective range of scores
would be compressed to 3 points. The well-known 20-point scoring
system developed at the University of California at Davis takes this
factor into account, and most individual judges' scores range from
about 9 to 19, with the average being close to 14. When the scores
given by individual judges on a panel are averaged, the range tends to
be further compressed (from about 11 to 17). With skilled judging
panels, difference of one or more points between two average scores
is probably significant, but a difference of less than a point is probably
not.

Two types of errors can obviously be made: awarding medals to
wines that do not deserve them, and not awarding medals to deserv-
ing wines. The author's mathematical study of this problem showed
that larger judging panels are less likely to commit the first type of
error and going by a majority opinion rather than demanding a consensus tends to reduce the second type. In the author's experience at
several state fair wine competitions, basing awards on a majority vote
rather than requiring unanimity seems to increase the number of
awards (especially the lower ones). Unfortunately, this tends to make
both winemakers and observers of wine competitions unhappy, the
former because they place little value on a bronze medal or an honorable mention, the latter because if 60–80% of all entered wines win an
award the whole competition can seem suspect. One partial answer
might be to eliminate bronze and honorable mention awards. Skilled
panels may be able to decide on their own how many awards are
appropriate. With less skilled panels it is sometimes desirable to set

an arbitrary predetermined number of awards and use a scoring system.

The overall quality of the wines and the purposes of the competition will obviously be major factors in deciding on the proportion of awards. Some amateur competitions limit awards to just one gold, silver, and bronze medal in each category, but that is usually a mistake. If the wine quality is fairly high, more awards may be in order. From the standpoint of educating winemakers or informing consumers, it is preferable that a given award in one category be equal to the same award in another.

EXAMPLE 18C
Selecting wine awards by ranking scores

In the 1979 Michigan State Fair Wine Competition, 4 panels of judges rated a total of 122 wines using a 20-point scale. Average scores from each panel were gathered and ranked from highest (17.5) to lowest (5.25). It was arbitrarily decided to give gold medals to the top 10% of the wines, silver medals to the next 10%, and bronze medals to the next 10%, but not to use a break point where scores were too close together. Under this system, 11 wines won gold medals (scores from 17.5 to 15.75), 11 won silver (15.25 to 14.2), and 13 won bronze (14.0 to 13.4).

When the gold-medal wines were rejudged by the entire panel to give a best-of-show award, average scores ranged from 17.00 to 14.31. This is how the top 4 scores looked:

Average score	Range of individual scores
17.00	19–15
16.69	19–15
16.62	20–15
16.46	19–15

From these results it is difficult to claim that the best-of-show wine was any better than the three runners-up.

Some competition neophytes place great stock in a particular judging system. In truth, this is really a secondary matter. If judges are familiar with the wines, the outcome will generally be the same whether they rank them or give them scores. Using scores can speed up a judging because judges do not have to confer. On the other hand, some judges prefer deciding directly on the awards each wine will receive because they have more of a feeling of control than they have with a scoring system.

Maynard Amerine and Edward Roessler, two professors retired from the University of California at Davis, have described many of the factors that can affect sensory response to wines. The sensitivity of judges to various wine constituents varies greatly. For example, about 25% of men are color blind to some degree. Motivation, concentration, and memory are all important in the judging process.

Judges can make a number of different errors that are psychological. A common one is the "time-order" error: preferring the first wine served. Another is the "contrast" error: overreacting to a good wine following a bad one and vice versa. The "stimulus" error occurs when judges use irrelevant criteria—the type of bottle closure, for example—when rating a wine. This type of error can also be caused by reaction to evaluations of other judges on the same panel. Judges can also make logical errors, assuming, for example, that one characteristic—e.g., a brown color—must necessarily mean that a wine has another characteristic—e.g., is very old or oxidized. Or judges can be overly lenient in evaluating a wine they believe to have been produced by a winemaker they know and like. It is also possible when using scoring scales for a judge to be influenced by a high (or low) score given to one characteristic into giving a high (or low) score to other characteristics. An error in association occurs with a series of wines when judges tend to give the same scores in succession, perhaps because of uncertainty or boredom. Finally, an uncertain judge may play safe by giving a mid-range score.

Interpreting the Results of Judgings

People often misunderstand what can and cannot be accomplished in a competition. When the results do not come out the way they were "supposed to," observers sometimes charge a judging panel with incompetence or even dishonesty. While some inexperienced judges make gross errors, in the better competitions this is not a major problem.

What judges do, consciously or subconsciously, is to compare the wine before them with their memories of the best and worst wines of the same type that they have tasted, placing the present wine somewhere on a mental scale from good to bad. Apparently judges faced with unfamiliar wine types tend to be more generous than they are with types they know well. Thus panels sometimes give gold medals to labrusca wines and only silver medals to vinifera wines, although in circumstances other than the competition the judges would prefer the vinifera.

"Highly trained" judges may reach a good agreement on a limited range of wines in competition after competition. Training means in part that the judges have agreed among themselves to approve of certain wine characteristics. No one is born with an infallible ability to judge wines, and most odor and flavor preferences are learned. Different trained judging panels often rank the same wines quite differently. Agreement among panel members does not ensure that their judgments are "correct." But their consistency in making judgments enables those entering wines to know where they stand.

Panels that are not highly trained often have major disagreements. If a few panel members change from year to year, results may well be inconsistent, leading some competitors to feel cheated. Competition entrants should recognize the limitations of human judgments and not be overly elated or depressed by one judging. A winemaker who enters tough competitions and has a batting average above 0.300 is doing well.

Much fuss is made over gold medals, and less attention is paid to silver and bronze. In fact there is often little statistical significance between them. The fairness of judging may be even more in question when a "best-of-show" award is given, the implication being that one wine is clearly superior to all the rest. This is seldom the case and frequently the voting is very close.

Wine competitions are not meaningless, as some observers complain, but should be viewed as imprecise measuring instruments. A crude measurement is usually better than none at all, and wine judging is improving in precision. Competition results are useful for indicating trends, including shifts in grape or fruit varieties used, in winemaking styles, and in overall quality from regions or individual winemakers. In eastern North America, for example, recent competition results clearly indicate that Canadian winemakers are rapidly improving their wines.

Hints on Winning Competitions

There are several tactics that an entrant can use to increase his or her chances of winning an award in a wine competition. Judges often react unfavorably to wines that differ from what they expect in any category. Entrants should become familiar with judges' biases in past competitions and enter their wines in categories where they will avoid such biases. The entrant's goal should be to make the "cut" and stay in the competition until the final round. If, for example, panels in a certain competition have historically favored dry vinifera table wines, it could be a mistake to enter a sweet Chardonnay in that category. Entering it in a dessert or miscellaneous category would probably improve its chances of winning. If rules permit, it is usually advantageous to enter the same wine in more than one category.

One can learn a lot by entering as many wines as one has available—if rules permit. Winemakers tend to favor wines they have labored the hardest to produce and to undervalue those that were easy to make. By entering as many wines as possible, a winemaker can gain a better appreciation of how his or her wines compare with others of the same type and may win some unexpected awards.

Wine Label Competitions

Wine label competitions are a rather new arena for amateur winemakers. In the mid-1970s the author ran what was probably the first national amateur wine label competition. The winning label appeared on a magazine cover, and since that time wine label competitions have received more attention. The American Wine Society has held a national amateur wine label competition for the past few years and other groups are also showing interest.

Wine labels offer winemakers an opportunity to display their originality as surely as they can in their wines. We have reached the point where the best amateur wine label designers are on a par with professionals and both can learn from each other.

19

Wine Storage and Record Keeping

Storage Containers

A generation ago it was not uncommon to find families who would make a barrel of wine and then tap it to drink, little by little. This primitive method almost guaranteed that the wine would become progressively more oxidized or vinegary as the barrel was slowly emptied; it does such violence to wine quality that it must be condemned.

Ordinary wines meant for quick consumption by families can be stored in gallon jugs and will keep fairly well in a partly full jug for a day or two after opening. After that, wine can be better preserved if transferred to smaller bottles. White jug wines may benefit from refrigeration after opening.

Better wines benefit from bottle aging, which proceeds faster in smaller bottles, and because the best wines are often in short supply they must be savored in small quantities. For most wine consumers the best-sized storage container is the 750-ml bottle (formerly known as the "fifth"), which holds an amount that can be drunk by two people during a meal. Even a single person can plan meals to use up a bottle in 2 days' time. A major advantage of the standard 750-ml bottle is that most wine racks are designed for it. Most available information on lengths of time various types of wine should be bottle-

| Bordeaux | German | Burgundy |

Standard bottle types

aged assumes this size bottle. Its three main shapes are the Bordeaux, the Burgundy, and the German. Darker glass protects wines from light better than clear glass.

The 375-ml bottle has some advantages. Not only is it handy for picnics, but red wines may age more quickly in the smaller size; and it is more practical for very rich, special, or unusual wines intended to be consumed in small quantities. For various practical reasons, most small winemakers will want to avoid bottles of other sizes.

Commercial wineries are now experimenting with bag-in-a-box containers, metal cans, and other nonconventional packaging.

Seals for Storage Containers

Screw-cap bottles have some advantages. Commercial wineries using screw caps employ machines that properly tighten them and either clinch them on the bottle or cover them with plastic sleeves. Amateur winemakers who reuse these caps save the trouble and expense of corking, but risk having the caps loosen during storage. Screw caps may also rust, but their main disadvantage is that consumers associate them with cheap wines.

Corks are the traditional seal for fine wines and, as such, appeal to consumers. They provide no special aging benefit (contrary to what was once believed). Sound corks can protect a wine for up to 20 years

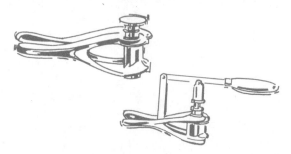

Hand corking devices

and, of course, will not rust. Unfortunately, cork quality seems to be declining and corks that leak are becoming more common. Minor leakage is usually a cosmetic defect, however, and of more concern to commercial wineries than to amateur winemakers. Bottles sealed with corks require special storage. Because corks must be kept wet to prevent shrinkage that would allow air to get into the wine, bottles must be stored either on their sides or upside down.

Amateurs winemakers not overly concerned with appearance or storage of bottles in bins can bottle still wines in used pop or beer bottles closed with crown caps. Many different bottle sizes are available, crown-capped bottles need not be stored on their sides, and crown caps will remain airtight for an indefinite period.

Labels and Capsules on Wine Bottles

The label plays an important role in the presentation of a bottle of wine. A well-designed label increases the enthusiasm of consumers and provides useful information.

For everyday wines it is sufficient for the amateur's label to tell the grape or other ingredients, the year of harvest, and any other information of interest. Gummed labels are preferable to self-adhesive because they are easier to remove when the bottles are recycled.

For special wines that are to be served to guests, the amateur winemaker may want a more decorative label. Most stores selling home winemaking supplies carry a variety of labels. Numerous home winemakers have designed their own, and a few have invested in presses to make them. Simple designs and basic colors (if colors are used) are usually most effective, and the decorative motif and necessary information should be in suitable proportion to each other. Some suppliers will make labels to order.

BEACHAVEN RED

Grown in Tennessee

Michigan

Vidal

A dry, golden grape wine with the characteristic taste and bouquet of a unique white from France. Grown, produced and bottled at

HAPPY WOODS VINEYARDS

By Dick, Glorie, Heidi and Mike Herrmann

ST. JOSEPH, MICHIGAN, U.S.A.　　ALCOHOL 12½% by Volume

Vintage　　Bottle Number_____
Of a total of_____

1982

TRAVILAH WHITE

SEYVAL

this harvest produced
26 *bottles*
of which this is
no. 16

from vines at bonlee produced by James H. Henry

BODENDORFER

" ZUR LIEBE DES WEINTRINKIN "

MADE FROM MICHIGAN GRAPES
TOM BUNDORF
FARMINGTON HILLS MI 48018

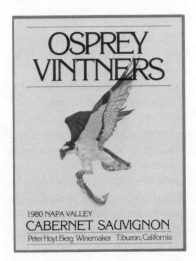

OSPREY VINTNERS

1980 NAPA VALLEY
CABERNET SAUVIGNON
Peter Hoyt Berg Winemaker Tiburon, California

Capsules to cover the tops of wine bottles can give a finished and professional appearance. They hide minor leakage around corks, keep corks clean, and can reinforce the label motif. The cheapest ones available are colored aluminum foil. Unfortunately, these tend to look amateurish because they cannot be completely smoothed out. More expensive and elegant foils made from lead are used by commercial wineries around the world. Amateur winemakers can get a fairly smooth appearance with lead foils by using a simple push-on device with a rubber doughnut inside. Small commercial wineries can get a better effect with a more expensive tool rotated by a motor.

Various plastic capsules are available. Some, made from heavy plastic, are meant to be pushed over a cork to fit the neck. Because amateur winemakers usually have a collection of many different bottles, such capsules do not always fit well. Others are made to be stored wet and shrink when they are placed on a bottle. These work well if used soon, but they are subject to mold attack if stored very long.

Probably the most suitable capsules for many amateur and small commercial winemakers are the heat-shrinkable type. These are inexpensive, can be stored indefinitely before use, fit a variety of bottle necks, and have a neat appearance. They can be heated and shrunk with a heat gun (similar to a hair drier but putting out hotter air) or a heat chamber specially designed for this application.

Storage Racks

Many wine storage schemes are possible. The main features of a suitable storage area are: (1) fairly constant temperature in the range of 55–70° F, (2) darkness, (3) absence of vibration, and (4) security against unauthorized wine removal. Serious storage areas should be big enough to hold both wines for current consumption and wines being bottle aged.

The simplest storage area is a bin with two side walls, a type sometimes used by small commercial wineries that store wines for aging and often do not label them until they are ready for sale. In these bins wine bottles are usually stacked in double rows with the necks facing each other. It is impossible to remove bottles except from the top layer. Such an arrangement is not suitable for amateur winemakers unless they have a large quantity of only a few types of wines.

The simplest bin for storing different types of wines so that they can be removed easily is a structure, usually attached to a wall, that has a series of small compartments, either square or diamond-shaped, each a foot or two in width and height. Such bins, suitable for the amateur, are also used by small commercial wineries in their tasting rooms.

When many different wines are to be stored, one useful arrangement can be made from a large number of plastic drain pipes fitted together and glued or otherwise fixed to maintain a stable structure. Pipes 4 inches in diameter, cut into 12-inch lengths, are suitable, and one bottle can be placed in each pipe. This type of storage is more work to construct and takes up more room than some others but can be quite decorative; the pipes can be painted.

A practical type of storage bin for the home winemaker who has several hundred bottles is a metal cage, sometimes called a "wine jail." Some of these are made in France, and they come in 100-, 200-, and 300-bottle sizes. These are often used in restaurants because each bottle can be viewed, it is easy to remove one bottle without disturbing the others, and the cages can be locked. This is a very compact form of storage and the racks fold up for easy transportation. The initial investment, however, is sizable.

Most of the numerous small wine racks on the market that hold one or several cases of wine are very expensive per unit of storage and are designed principally as decorations. For a person who has limited space, storing wine on shelves in the back of a closet is perfectly feasible, provided that the temperature does not vary widely with the seasons. Wines can be stored in cardboard or wood cases. If shelves are fitted with dividers, individual bottles can be binned.

For those who live in warm climates and need a cool storage area, several types of refrigerated wine vaults are on the market. These are fairly expensive but are self-contained and do not require any construction. Someone who has a basement can construct a refrigerated room using insulation and a small window air conditioner controlled with a thermostat. For best results, the thermostat should be placed at the end of the room farthest from the air conditioner.

The Need for Wine Records

Just as winemakers store wine in a manner that keeps it safe and accessible to them, so should they keep information. No vineyardist or winemaker can afford to be without accurate records. Successful vineyard management requires records that show what has been done and how the vines responded. Successful winemaking requires records for each wine, showing treatments received, development, and sensory evaluation. Vineyardists and winemakers also need information files on topics that can help them improve their practices (e.g. manufacturers' literature, magazine articles).

There is no best information-recording system. Some people need only simple records, others require a system capable of accurately recording and cross-indexing vast amounts of data. The main requirement of a record system is that it serve its intended purpose with a minimum of effort. A system too burdensome to keep up is as bad as no system at all.

Vineyard Records

Properly managing a vineyard requires keeping records in the following categories:

(1) Pruning severity (wood removed and buds left per vine), fertilizer applications, irrigation, weed control measures, sprays applied, tying up or positioning of vine shoots, bird and other pest control measures, post-harvest treatments, and the dates of all treatments.

(2) Weather information, including dates of the last frost in the spring and the first in the autumn, average temperatures and temperature extremes, and hail and other special events.

(3) Grape yields and grape and wine quality for the smallest practical vineyard units. (This information can identify problem areas within a vineyard.)

(4) Information on individual vines, including crop size and health. (Vines that bear smaller-than-usual crops should be marked for future pruning adjustments. Dead vines should be noted so replacements can be planned.)

(5) Costs of materials and labor, preferably by grape variety. (Such information helps to justify continuing or expanding a vineyard.)

Two useful tools in vineyard record keeping are a small tape recorder and vineyard maps. A pocket-size tape recorder is ideal for recording information when walking through or working in a vineyard. A map of all individual vines, showing grape variety and general condition of each vine, is most valuable. Other maps can show larger features of the vineyard, such as low spots and heavy soils.

An example of a profit analysis for a small family vineyard is given below.

EXAMPLE 19A
Vineyard profit analysis

Income

Value of crop retained (1100 lb @ $0.45/lb)	$495.00
Value of crop sold (600 lb @ $0.45/lb)	$270.00
	$765.00

Expenses

Equipment and supplies	$105.00
Cultivating contracted for	$ 75.00
Hired labor	$ 45.00
Auto expenses to and from vineyard (275 miles @ $0.20/mile)	$ 55.00
	$280.00
Gross profit	$485.00
Hours family worked in vineyard	95 hr
Value of family labor per hour	$ 5.11

Winemaking Records

Categories of information that should be recorded in winemaking include:

(1) Grape properties, including sugar and acid (plus pH if possible).

(2) Crushing and pressing details, including presence of stems, pressing aids used, juice yield per weight of grapes, amount of free-run and press juice.

(3) Cellar treatments, including SO_2 additions, rackings, finings, filterings, and detartration.

(4) Individual barrel records, including dates of filling and emptying, the quality of the wine coming from the barrel, whether or not a malolactic fermentation occurred spontaneously in the barrel, the barrel's cleaning history, and its physical condition. It is important to know where each barrel is in its useful lifetime of providing extract (about 7 years). Barrels that have developed major problems (e.g., bad smells, leakage) should be identified and discarded as soon as possible.

(5) Chemical and sensory evaluations of the wines as they progress (helpful in revealing the efficacy of cellar treatments.)

It is worthwhile for a winemaker to keep track of the materials used and time spent on each batch of wine so that production costs can be calculated for each.

EXAMPLE 19B
Record of treatments, red wine

Date	Wine/must properties	Treatment
10-3-78	Brix 19, acid 1.10%	Stemmed & crushed, added 50 ppm SO$_2$, Champagne yeast
10-6-78	Brix 10	
10-9-78	Brix 4	Must pressed
10-21-78	Residual sugar 0.1%, acid 1.01%	Wine racked, ML culture added
11-20-78	Acid 0.78%	
11-30-78	Acid 0.76%	Wine racked, 50 ppm SO$_2$ & 1 g/gal bentonite added
12-14-78		Wine racked, cooled to 28 F
12-28-78	Acid 0.71%	Wine racked, 15 g/gal oak chips added
3-11-79		Wine racked, 30 ppm SO$_2$ added, wine bottled

Federal regulations require a commercial winery to keep many detailed records. Richard P. Vine's book, *Commercial Winemaking* describes them. Vine discusses wine lot numbering systems and a 3-digit code that can be used to identify batches of wines and their progress. An example of such a coding system is:
First digit shows vintage year:

> 0 = 1980, 1990, etc.
> 1 = 1981, 1991, etc.

Second digit shows origin of grapes:

> 0 = grown by winery on estate
> 1 = grown by winery on nearby property
> 2 = purchased locally

3 = purchased from same general area
4 = purchased from out-of-state

Third digit shows status of wines:

0 = fermenting must
1 = new wine after first racking
2 = new wine after second racking
3 = fining agents added
4 = wine in detartration
5 = wine being aged
6 = wine after blending
7 = wine ready for bottling

In this system, a code of "345 Chardonnay" denotes a 1983 Chardonnay wine made from grapes purchased from out-of-state that is presently being aged. Several wines with this same description can be identified by additional numbers: 145-1, 145-2, etc. No possible coding system will suit all winemakers, but individual modifications are easy to make.

Wine-Tasting Records

Serious winemakers will keep accurate records of tastings of their own wines and of other commercial wines. Both numerical and descriptive systems can be used.

Numerical wine notes can contain information on appearance, fragrance, taste, physical impressions (such as body or bubbles), technical flaws (oxidation, bad smells, volatile acidity) and overall quality. One approach is to assign numbers to each of these attributes, using a scale of 0 to 4 (where 4 is best). A code of 20123/8 would mean, for example: (1) average appearance, (2) much less than the expected amount of fragrance, (3) below average taste, (4) average body or other physical attributes, (5) only minor technical flaws, and (6) below average overall quality. (The 8 is the total score out of a possible 20.) For more detailed ratings the University of California at Davis 20-point system is popular.

Actual descriptions of wines are often more useful than numerical summaries. The wine coded above might be described as: medium red color, undesirable weedy fragrance, slightly herbaceous taste, average body, a hint of chemical aftertaste, overall a barely acceptable wine.

Measures, Conversion Factors, and Calculations

Most books on winemaking contain extensive tables. These take up room that could be put to better use and often imply need for a greater accuracy than is warranted (as when the amount of sugar to add to a must is given to four decimal places). No user of tables can be expected to commit them to memory and without them one is usually helpless. When the required values are not found in particular tables, interpolation or extrapolation becomes a burdensome necessity. Today, with inexpensive hand-held calculators widely available, it makes sense for conversions to be given as equations that can be easily worked that way.

Weights

The following abbreviations are commonly used:

mg = milligrams = 1/1000 gram
g = grams
kg = kilograms = 1000 grams
oz = ounces
lb = pounds

The following conversion factors are useful:

$$1 \text{ oz} = 28.4 \text{ g}$$
$$1 \text{ g} = 0.0353 \text{ oz}$$
$$1 \text{ lb} = 454 \text{ g}$$
$$1 \text{ kg} = 2.20 \text{ lb}$$

Volumes

The following abbreviations are commonly used:

ml = milliliters = 1/1000 liter
L = liters
hL = hectoliters = 100 liters
gal = gallons

The following conversion factors are useful:

1 fluid oz (U.S.) = 29.6 ml
1 gal (U.S.) = 3.785 L
1 hL = 264 gal (U.S.)
1 Imperial gallon = 1.2 U.S. gallons
1 Imperial fluid ounce = 0.96 U.S. fluid ounce

Dry Volume Weight Estimates

The following list gives a rough estimate of the volume of various dry substances per unit of weight. The user should appreciate that errors of 10–20% are possible depending on how finely divided and tightly packed these substances are.

Cane sugar	2.5 cups/lb
Potassium metabisulfite	5.5 g/tsp
Sodium bisulfite	5.5 g/tsp
Tartaric acid	5.5 g/tsp
Bentonite	5.2 g/tsp
Citric acid	4.6 g/tsp
Pectic enzyme powder	4.0 g/tsp
Calcium carbonate	4.0 g/tsp
Gelatin crystals	3.3 g/tsp
Decolorizing charcoal	1.4 g/tsp
Tannic acid	1.3 g/tsp
Sparkolloid®	1.3 g/tsp

Soluble Solids Conversions

A variety of hydrometer scales are used to indicate soluble solids in grape and other musts. The specific gravity of a liquid is the ratio of the weight of the liquid to the weight of an equal volume of water at the same temperature. In Germany the Oeschle scale is used and is defined as:

$$°Oeschle = (sp.~gr. - 1) \times 1000$$

Amateur winemaking literature often uses the term "gravity" and this is identical to °Oeschle.

Many hydrometers are calibrated in °Brix. The definition of °Brix is:

$$°Brix = g~soluble~solids/100~g~of~solution$$

The relationship between the Brix and Oeschle scales (for normal musts) is given by:

$$°Brix = (0.22 \times °Oeschle) + 1.6$$

Fermentable sugars make up most but not all of the soluble solids in a must. The following relationships estimate the sugar in a must:

$$\%~sugar = (°Brix - 3) \times specific~gravity$$

$$\%~sugar = (°Oeschle/4) - 2.5$$

If a hydrometer is used to measure the °Brix of a must that is not at the temperature for which the hydrometer was calibrated (usually 60° F), one should subtract 0.03° Brix for each °F below the calibration temperature and add 0.03° Brix for each °F above.

Sugar-to-Alcohol Conversions

The theoretical yield of alcohol is 51.1% (by weight) of the sugar content of a must. In practice the yield is somewhat lower. The following relationship is based on average experience.

$$\%~(by~volume)~alcohol = °Brix \times 0.59$$

For red grapes from hot areas the relationship is more nearly:

$$\%~(by~volume)~alcohol = °Brix \times 0.54$$

For most estimations the following relationship is adequate:

$$\% \text{ (by volume) alcohol} = °\text{Oeschle}/8$$

Sugar Additions to Musts

For grape musts that are slightly deficient in sugar many wine-makers add cane sugar. For moderate corrections, one can use the rule that 1.5 oz of sugar will raise a gallon of must by 1° Brix.

Extract Estimation in Wines

The extract of wines is the soluble solids. In dry wines it is made up of acids, minerals, tannings, unfermentable sugars, and so forth. In sweet wines it also includes fermentable sugars. The extract content of most dry grape wines is 2–2.5%. Extract content can be estimated if one knows the alcohol content and the °Brix of a wine. For table wines the following approximate relationship applies:

$$\% \text{ extract} = (\% \text{ alcohol} \times 0.3) + °\text{Brix} + 0.5$$

For dessert wines the approximate relationship is:

$$\% \text{ extract} = (\% \text{ alcohol} \times 0.25) + °\text{Brix} + 1.3$$

Temperature corrections for hydrometers vary with alcohol content. A user is better advised to adjust the temperature of the wine to the hydrometer's calibration temperature than to fool with temperature corrections.

Alcohol Conversions

For the range of alcohol found in wines the following relationships are sufficiently accurate:

Volume % alcohol = 1.25 × weight % alcohol

Sp. gr. decrease = 0.0011 × (% alcohol increase)

When the alcohol content of off-dry wines is measured by ebulliometer, a better estimate of true alcohol content can be obtained by subtracting 0.05% alcohol for each 1% sugar in the wine.

Calculating Dosages

When treating wines with chemicals, the following information is useful:

$$1 \text{ mg/L} = 1 \text{ ppm}$$
$$1 \text{ g/L} = 0.1\% = 1000 \text{ ppm}$$
$$1 \text{ g/gal} = 0.026\% \text{ or } 260 \text{ ppm}$$
$$1 \text{ lb/1000 gal} = 120 \text{ mg/L} = 120 \text{ ppm}$$
$$1 \text{ g/L} = 8.3 \text{ lb/1000 gal}$$

For treating wine with sulfur dioxide or a variety of fining agents it is usually easiest to prepare a 3–10% solution and add it to the wine. For example, if one does fining trials and finds that 20 ml/L of a 5% bentonite suspension is the proper amount, this is equivalent to 1 g/L of dry bentonite or 8.3 lb/1000 gal. To treat a 50-gallon barrel of wine, one would need 8.3 lb × 50 /1000 or 0.42 lb of bentonite.

When one uses potassium metabisulfite or sodium bisulfite, slightly over half the weight will be sulfur dioxide. These dosage rates are reasonable approximations:

$$1 \text{ g/L of bisulfite} = 600 \text{ ppm}$$
$$1 \text{ g/gal of bisulfite} = 150 \text{ ppm}$$
$$1 \text{ ml/gal of 5\% bisulfite solution} = 7.5 \text{ ppm}$$

It is possible to calculate dosage rates more accurately than this, but what is gained? Chemicals are never 100% pure and volumes are seldom known exactly. It is preferable to use the approximate factors given above and accept a 10% error in dosage rate rather than get into involved calculations that can lead to a tenfold error if a decimal point is displaced.

Temperature Conversions

The temperature system most used in the United States is the Fahrenheit system, in which water freezes at 32° and boils at 212°. Most of the rest of the world uses the Celsius system, in which water freezes at 0° and boils at 100°. The number of Fahrenheit degrees is 1.8 times the number of Celsius degrees for any temperature range. The following formulas allow for an easily remembered conversion: To convert from °C to °F:

Add 40 to the °C; multiply by 1.8; subtract 40.

To convert from °F to °C:

Add 40 to the °F; divide by 1.8; subtract 40.

These formulas are easy to remember because one always adds 40 at the beginning and subtracts 40 at the end. In °F there are more degrees so one multiplies by 1.8; in °C there are fewer, so one divides by 1.8.

Laboratory Evaluation of Wines

Laboratory analyses of wines are routinely done by large wineries for daily quality control and to maintain consistent quality year after year. Analyses supplement sensory evaluation in monitoring wine quality and are very useful when a winemaker is producing a new type of wine or runs into problems.

A wide variety of important analyses can be done with relatively inexpensive laboratory equipment. Listed here are those likely to fall within the scope of home winemakers' equipment and technique. References mentioned in the text are listed at the end of this appendix.

Soluble Solids

Soluble solids (mostly sugar) in grape juice, fermenting must, and wine are usually measured with a hydrometer (about $5) or refractometer (about $100) and reported as °Brix. Alcohol in musts and wines affects readings of hydrometers and refractometers differently. A 12% alcohol in water solution gives a −4° Brix reading with a hydrometer but a +4° Brix reading with a refractometer. Extract contributes about 2° Brix, so a dry wine gives about a −2° Brix reading with a hydrometer and a +6° Brix reading with a refractometer. Readings significantly higher than this are usually due to sugar and can be used to estimate the sugar in sweet wines.

Alcohol

Most small wineries measure wine alcohol with an ebulliometer (about $175), which is accurate to about 0.2% with dry wines. With slightly sweet wines, a correction needs to be made as detailed by Amerine and Ough. With sweet wines, more accuracy can be achieved by making a distillation prior to the alcohol determination. A dichromate titration is sometimes used to measure alcohol, and Amerine and Ough give directions. The author has tried several inexpensive glass ebulliometers but found them to be very inaccurate, probably due to poor design. Professional quality ebulliometers are made from metal.

Amateur winemakers sometimes rely upon a vinometer ($2–3) for alcohol readings. This little capillary tube device can give readings accurate to about 1% if the tube is carefully cleaned before use, if several determinations are made and averaged, and if the vinometer is calibrated with wines of known alcohol strength.

Extract

Measuring wine extract is easily done using a laboratory hydrometer (about $10) and applying a correction for alcohol content. Amerine and Ough give details. An extract measurement is often useful with wines made from native American grapes or with nongrape wines where water is added and the extract is low. By combining extract measurement and sensory evaluation, a winemaker can decide on the amount of sugar that will give the desired body to low-extract wines.

Residual Sugar

A quick qualitative test for residual sugar can be made with products sold in drug stores to measure urine sugar such as Clinistix® test strips. Winemakers should check wines after the first or second racking to be sure that fermentation is completed.

Semiquantitative sugar measurements on slightly sweet wines can be made with the Dextrocheck® kits sold by home winemaking shops. This simple test measures sugar up to 2% (or more if the wine is diluted before the test).

For quantitative sugar determinations, a variety of procedures based on Fehling's reagents are used. Amerine and Ough detail several methods. The method of Mauer, referenced by Amerine and Ough, is relatively simple and could be useful for winemakers doing routine sugar analyses. It requires a colorimeter but is rapid and needs fewer chemical solutions than most other methods.

Glucose and Fructose

Some wineries measure glucose and fructose in almost dry wines by an accurate enzymatic method. The 1978 McCloskey reference gives details of a rapid method.

Total Acids

Total acidity is determined by a titration where a base (usually sodium hydroxide) is slowly added to a wine until an indicator (usually phenolphthalein) changes color or until the solution reaches a pH of 8.2. Dissolved carbon dioxide in a must or a wine gives a false endpoint to a titration, so it is essential to heat all must or wine samples just to a boil—expelling carbon dioxide—before the titration. Such heating can be conveniently done in a microwave oven.

Most home winemaking supply shops sell simple acid determining kits (about $10) which use a syringe to dispense the sodium hydroxide solution. More accuracy is obtained by using a burette. Amerine and Ough, and Vine give details of acid titrations. Devices that use the evolution of carbon dioxide in a tube to measure acidity are much less reliable than a titration.

Small wineries and home winemakers who do not want to be bothered with preparing solutions might be interested in the digital titrator system sold by Hach (about $100). This system uses small, sealed tubes of concentrated titrants and does not require the setup and cleanup that a burette titration requires. The system can also be used for volatile acidity, sulfur dioxide, and other determinations.

pH

The only practical way to measure the pH of grape musts and wines accurately is with an electronic pH meter. Many models are available, starting at about $100. These meters are also useful in determining the end points of acid titrations. Both dark red wines and the dilute solutions used in volatile acidity titrations make an accurate visual end point determination difficult. Even unskilled or color-blind analysts can get reliable results with a pH meter.

A saturated solution of potassium bitartrate (cream of tartar) in water is useful for standardizing a pH meter. At room temperature this solution has a pH of 3.55. It spoils easily so it should be used within a few weeks.

Volatile Acidity

Commercial wineries measure volatile acidity because government regulations limit this component. A Cash still ($150–200) is commonly used. Home winemakers may be able to construct a less expensive apparatus based on a Sellier tube, mentioned in Amerine and Ough.

Total Phenols

Phenolic wine components include red pigments, tannins (usually the major phenolic), and related substances. Individual phenolics are very difficult to measure so a measurement of total phenols is often used to estimate tannins. The usual method employs a Folin-Ciocalteau reagent and a colorimeter, detailed by Amerine and Ough.

Sulfur Dioxide

Free sulfur dioxide is commonly determined by an iodine titration in the Ripper method. Because the titration end point is difficult to determine with red wines, an aeration-oxidation method is often preferable for them. Total sulfur dioxide is determined by the same methods after a basic hydrolysis with sodium hydroxide. Amerine and Ough give details.

Winemakers can avoid standardizing the iodine solution and simplify the process by using the Hach digital titration system. They might also try a platinum electrode to determine the end point (see *The Practical Winery* reference).

Tartaric Acid

Tartaric acid is usually the major acid in grapes and wines. Its measurement can be important when one is using the double-salt method of acid reduction or dealing with wines of normal total acidity but high pH. Tartaric acid is conveniently measured by means of the vanadate procedure and a colorimeter. Amerine and Ough supply details. The wine sample can be decolorized with a short column of PolyClar AT® before the determination. This procedure is superior to the older, messier one of boiling wine with decolorizing charcoal. The Mattick and Rice paper explains the procedure.

Malic Acid

Malic acid is usually determined qualitatively by paper chromatography in order to follow the course of a malolactic fermentation. Amerine and Ough give details. A newer method, which Boulton describes, uses a homemade electrode with a pH meter. Enzymatic methods may also interest small wineries and McCloskey (1980) explains them.

Color

For side-by-side comparisons of wine color, the human eye is often more sensitive than laboratory instruments. But side-by-side comparisons are not feasible when one is attempting to match the darkness of a red wine of which one has no sample. Even if a standard wine is available, its color may have changed with age. A sizable portion of the male population is color blind, which makes accurate color matching difficult or impossible for them. In such cases, a colorimeter is valuable. The ratio of absorbance at 420 nanometers (nm) to absorbance at 520 nm estimates browning of red wines, and the sum of the absorbances at these two wavelengths estimates the depth of wine color. Details and further references are given by Amerine and Ough.

Haze

A quantitative measurement of wine haze can be best made with a ratio nepholometer (about $1000) that can compensate for wine color. A semi-quantitative measure of haze can be made by comparing in a colorimeter the light transmittance of an untreated wine sample with the same wine passed through a 0.2-micron filter. The Antlia hand pump filter system from Schleicher & Schuell (about $200) is a convenient way to clarify small wine samples for this and other purposes (it is sold by VWR; see Appendix D).

The clarity of white wines can be roughly estimated by viewing printed matter down through 50 ml cylinders filled with wine. Another rough method uses the Tyndall effect. If a penlight is used to shine a beam of light through a glass container of wine in a darkened room, brilliant wines will not show an outline of the light beam in the wine but less clear wines will show a definite light pathway.

Electrical Conductivity

Wine conductivity can be measured with an electronic conductivity meter ($200 and up). Suitable meters have several conductivity ranges (at least 1000 to 5000 micromhos/cm) and temperature compensation circuits. Conductivity measurements are very useful in monitoring the cold stabilization of wines. Dry red wines are considered cold-stable at a conductivity of 1800 micromhos/cm or less, dry white wines at 1400–1600 micromhos/cm or less, and dessert wines at 1000–1200 micromhos/cm or less. Winemakers can use a conductivity meter to determine when wines have received adequate cold stabilization or in comparing the effectiveness of several methods of accomplishing it.

Sorbic Acid

Sorbic acid or its potassium salt is used in some sweet wines to prevent yeast growth. One determination involves thiobarbituric acid and a colorimeter. Amerine and Ough give details.

Total Cations

Measurement of total cations (hydrogen, potassium, sodium, calcium, and so forth) can be used as a check on other analyses or as a way to monitor the decrease in potassium during cold stabilization. To measure total cations, a wine sample is passed through a strongly acidic ion exchange column (such as Amberlite IR-120 AR), which exchanges hydrogen ions for cations. The treated wine is titrated and the moles of increased acidity equal the total nonhydrogen cations in the wine.

Potassium

The main reason to measure potassium in wines is to watch the progress of cold stabilization. Flame photometry or atomic absorption spectrophotometry is preferred by larger winery labs but both are impractical for most small wineries. The earlier Amerine lab manual details a simple precipitation and titration, which may be of use to small winemakers.

Calcium

Calcium analysis of wines is useful after addition of calcium carbonate, or the proprietary compounds Acidex® and Koldone®, for acid reduction and cold stabilization. Calcium tartrate tends to precipitate slowly, so monitoring calcium levels is necessary if winemakers do not want calcium deposits to form in bottled wines. Satisfactory methods of determining calcium use an ion-selective calcium electrode or a titration with EDTA. Amerine and Ough give details. A recent paper by Ough et al. describes a rapid colorimetric method.

Iron

Though iron is rare in modern wines, it can give hazes from ferric phosphate. Its presence can be measured with a colorimetric procedure detailed by Amerine and Ough. Wineries using well water to wash equipment should check the water's iron content to be sure that it is not abnormally high.

Dissolved Oxygen

Meters are available that can measure oxygen dissolved in wine. Some suppliers sell less expensive add-on equipment that is used with a pH meter. Some wineries measure dissolved oxygen frequently to monitor cellar operations and prevent excessive oxygen pickup. Vine and also Amerine and Ough discuss these procedures.

Identification of Sediments

Among the nonbiological sediments found in wine are lint, cellulose fiber, cork dust, tartrates, paraffin, protein, and tannins. Quinsland outlines identification procedures using a low-power microscrope and simple chemical reagents.

Test for Pectins

To determine if pectins are present in grape juice, mix equal parts of juice and denatured alcohol, shake the mixture, and see if a haze forms. Pectins are less soluble in alcohol than in water so haze above that in the original juice suggests their presence.

Cold Stability

There are several tests for cold stability. One is to put a small container of wine in a freezer, freeze it solid, thaw it, then check for a tartrate precipitate. The absence of a precipitate is reasonable evidence that the wine is cold-stable. Another method is to store a wine for 2 days at 28° F and check for tartrates.

Heat Stability

One method of determining wine heat stability is to heat a sample to boiling, cool it, and see if any flocculant precipitate is present. A variety of other empirical methods are used which involve heating wines for a longer time at a lower temperature, as for 2 days at 120° F.

Proteins in wine—generally associated with heat instability—can be qualitatively determined by adding 1 ml of 50% trichloroacetic acid to 10 ml of wine in a test tube, heating in boiling water for 2 minutes, then checking for clarity. Complete absence of a haze or deposit signals the absence of proteins.

Fining Trials

Effective use of fining agents requires laboratory tests before the bulk of the wine is fined. One method is to set up a series of small containers with 50 ml of wine and add different amounts of fining agent to each. For a fining agent in a 5% suspension or solution, 0.25 ml per 50 ml of wine equals approximately 1 gram per gallon. After these wine samples have been allowed to stand overnight, one should examine them for clarity, taste, and amount of sediment and then decide on the minimum fining agent to use on the bulk wine to get the desired effect.

Fermentation Tube

A fermentation tube is a small J-shaped container, open on the short side and sometimes having markings on the long side. A sample of wine is placed in the tube and inverted so that all air is expelled from the closed side of the tube. It is then left to stand at room temperature and observed for evidence of gas formation. This apparatus (less than $10) is valuable for detecting the onset of slow fermentations, verifying that fermentations are stuck, or checking the fermentability of still wines destined to be made into sparkling wines.

Microbiological Tests

A number of reasonably simple procedures can be used to monitor yeasts and other microorganisms in wines. Vine details some of these.

REFERENCES

M. A. Amerine, *Laboratory Procedures for Enologists*. University of California, Davis, 1965.

M. A. Amerine and C. S. Ough, *Methods for Analysis of Musts and Wines*. New York: Wiley, 1980.

Roger Boulton, *Am. J. Enol. Vitic., 29,* 289–291 (1978).

A. Massel, *Applied Wine Chemistry and Technology,* London: Heidelberg Publishers, 1969.

Leonard R. Mattick and Andrew C. Rice, *Am. J. Enol. Vitic., 32,* 297–298 (1981).

Leo P. McCloskey, *Am. J. Enol. Vitic., 29,* 226–227 (1978).

Leo P. McCloskey, *Am. J. Enol. Vitic., 31,* 212–215 (1980).

C. S. Ough, A. Caputi, and M. Groat, *Am. J. Enol. Vitic., 30,* 58–60 (1979).

The Practical Winery, August/September 1980, p. 9.

D. Quinsland, *Am. J. Enol. Vitic., 29,* 70–71 (1978).

Richard P. Vine, *Commercial Winemaking: Processing and Controls*. Westport, Conn.: AVI Publishing, 1981.

APPENDIX C

Wine Vinegar

High-quality wine vinegars are not the same as spoiled wines. A good vinegar retains much of the quality of the starting wine and is suitable for food use.

In the making of wine vinegar, special bacteria (usually strains of *Acetobacter aceti*) convert alcohol to acetic acid. These strains differ from the acetic bacteria flora sometimes found in wines (which winemakers do their best to eliminate) and from those used to make cider vinegar. In wines acetic bacteria are usually found as a thin grey surface film or distributed throughout the wine. In apple juice (cider) or fermented apple juice (hard cider), the acetic bacteria form a heavy film called "vinegar mother."

To ensure good-quality vinegar, one should purchase a vinegar bacteria culture. In most cases the natural mixture of acetic bacteria found in spoiled wines is not suitable because these bacteria produce noticeable amounts of ethyl acetate, which has an odor reminiscent of lacquer thinner, a smell no one wants in a vinegar-and-oil salad dressing. Commercial bacteria cultures do not give this smell.

Important factors influencing the growth of vinegar bacteria are oxygen, alcohol, pH, temperature, sulfur dioxide, and prior heat treatment of the wine. Without oxygen, vinegar bacteria soon die, but given much oxygen and time, acetobacter convert alcohol to carbon dioxide and water rather than acetic acid. At alcohol levels above

13%, most acetic bacteria grow very poorly; at alcohol levels below 10% the less desirable ones can thrive and the result is a vinegar of reduced quality (often with noticeable ethyl acetate). Vinegar bacteria grow much better at pH 4.5 than at pH 3.2, but so do many other bacteria. Sulfur dioxide prevents the growth of acetic and other bacteria. Wines made from grape concentrates or hot pressed grapes and wines that have been pasteurized are very difficult to make into vinegar. It may be that the heating produces some substances that are toxic to acetic bacteria.

In commercial vinegar production, several processes are used. In the Orleans process, wine is diluted to about 10% alcohol, and ¼ as much fresh wine vinegar is added as an innoculum. A barrel, with a bung hole and several other holes drilled in the top (screened to keep out insects), is filled about ¾ full of this mixture and stored at 70–85° F. Approximately ¼ of the barrel contents is removed every 3 to 4 months and replaced with wine. This process allows for aging as well as acetification and usually produces a superior vinegar.

Here is an example of vinegar production based on the Orleans process. To make a simple red wine vinegar, select a sound red wine with a pleasing taste and smell, a pH of 3.4 to 3.6, and low free sulfur dioxide (below 10 ppm). Red wines that have undergone a clean malolactic fermentation have the advantage that unwanted lactic bacteria (which can give bad odors and flavors) will not be competing with the acetic bacteria.

Add water as necessary to reduce the alcohol content to 10–11%. Fresh tap water should not be used because it contains chlorine that kills bacteria; water that has been boiled and allowed to cool is suitable. Fill a clean container (bottle or barrel) about ¾ full with the wine, allowing plenty of surface for air contact. Add a good vinegar bacterial culture, purchased or obtained from someone else who is making good wine vinegar. Commercial wine vinegars do not make suitable starters because these have usually been pasteurized or sterile-filtered to remove bacteria.

Stopper the container with fresh cotton wool, which allows the passage of air but is very effective in trapping microorganisms and keeping out unwanted bacteria, yeasts, molds, and insects. Set it in a warm place (70–85° F), away from excessive light. Depending on the size of the container, acetification will usually be completed in 1–3 months. Formation of acetic acid can be monitored by titrating a vinegar sample. The theoretical yield from 10 volume % (8 weight %) alcohol is 10.4 weight % acetic acid. The actual acetic acid content will be lower because of alcohol evaporation and partial oxidation to carbon dioxide and water. Commercial wine vinegars usually contain 7–8% acetic acid, which makes them stronger than distilled white vinegar, with 5% acetic acid. Finished vinegar can be diluted with water to the desired strength. Oak chips can be added to vinegar produced in a glass container to obtain an aged-in-oak flavor.

After acetification is completed, remove ¼ to ¾ of the vinegar and replace it with wine to keep the cycle going. Before adding the new wine, one must evaluate the vinegar to be certain that unwanted bacteria have not produced bad smells or tastes. Wine vinegars are difficult to evaluate by tasting. One useful technique is to dip in a sugar cube and then taste the cube; the sweetness helps to mask the high acidity so that the vinegar's other qualities can be judged.

Freshly produced vinegar will usually be cloudy and can be clarified by fining, just like wine. Bentonite is a suitable fining agent; gelatin or casein plus tannin can also be used. Vinegar that is only slightly hazy can be clarified by filtration.

Vinegars usually improve upon aging for several months in well-filled containers without access to air. Before vinegar is bottled, any remaining acetic bacteria should be killed by a flash pasteurization or the addition of 150 ppm of sulfur dioxide to the bottle.

White wine vinegars can be made much the same as reds. One thing to watch for in selecting white wines to start with is high sulfur dioxide, which kills vinegar bacteria. If free sulfur dioxide is too high, some of it can be dispelled by racking and aerating the wine. An alternative is to add a little hydrogen peroxide to neutralize the sulfur dioxide. Half a milliliter of 3% hydrogen peroxide added to a 750-ml bottle of wine will neutralize 38 ppm of sulfur dioxide. Excessive hydrogen peroxide should be avoided because it can darken a white wine and degrade quality.

Wine vinegars can be blended like wines. If vinegars have slightly wrong colors, odors, or tastes, blending can be used to correct these flaws.

Vinegars should be stored as carefully as wines, avoiding light, heat, and exposure to air. While most wine vinegars improve with a few months of aging they do not improve indefinitely and should be used within a year or two.

While wine vinegars represent a great improvement over commercial distilled white vinegar for many uses, flavored vinegars give an extra special touch. Examples of flavorings are basil, tarragon, mint, rosemary, thyme, oregano, red pepper, curry powder, dill seed, celery seed, fennel seed, mustard seed, lavender flowers, shallots, and garlic cloves.

The basic process used in flavoring vinegars is infusion. One mixes a base vinegar with the flavoring agent, puts the mixture in a closed jar, and shakes it every day for 10–14 days. Herbs and some of the other flavoring agents tend to mute the sharpness of white and red wine vinegars and produce a more mellow-tasting product.

Personal preference will determine the types and amounts of flavoring agents to use. As a starting point one can use 2 cups of chopped fresh herbs per quart of vinegar or ½ cup of dried herb leaves. With spices or seeds, 1 tablespoon per quart of vinegar is about right. Fresh herbs contain more of the flavorful essential oils. With dried herbs and seeds it is best to heat the vinegar to a simmer (almost boiling) before pouring it over the flavoring agent. Aromatic seeds such as mustard or celery should be bruised or crushed before use. Bruising can be done with a pestle in a mortar or by tying the seeds in a bag and tapping the bag with a hammer.

After the infusion process, powdered spices need to be carefully filtered from the vinegar (a coffee filter or a paper towel will usually suffice). With seeds and other flavoring agents that give up most of their flavor to the vinegar, filtering is optional and the spent seeds can be left in as a decorative touch.

Seasonings can be blended, but one component (e.g., garlic) should not be used in an amount that will overpower the rest.

General Suppliers to Home and Small Commercial Winemakers

The small shops in almost every state that sell supplies to home winemakers are too numerous to list, but many of them belong to a trade organization (HWBTA) that can provide up-to-date information on the shops in each area: Home Wine and Beer Trade Association, c/o *Beverage Communicator,* Box 43, Hartsdale, N.Y. 10530.

Two suppliers have a particularly extensive retail and mail order business in supplies for home winemakers and small commercial wineries. These are: Presque Isle Wine Cellars, 9440 Buffalo Road, North East, Pa. 16428; and Wine and the People, 907 University Avenue, Berkeley, Cal. 94710. Both offer fresh grapes in season, and Wine and the People ships premium frozen grapes across the country.

A supplier of herbs for vermouth production is Aphrodisia, 21 Carmine Street, New York, N.Y. 10014.

Specialized Commercial Winemaking Suppliers

There are many hundreds of suppliers that offer equipment and supplies to commercial winemakers. Two annual directories that list many suppliers are: Directory Issue, *Eastern Grape Grower & Winery News,* Box 329, Watkins Glen, N.Y. 14891; and Buyer's Guide Issue, *Wines & Vines,* 1800 Lincoln Avenue, San Rafael, Cal. 94901.

Laboratories

Laboratories that specialize in analyzing wines often also offer yeast and malolactic bacteria cultures and certain lab supplies. They include:

Finer Filter Products,
Div. of Cellulo Co.,
6821 Central Avenue,
Newark, Cal. 94560;

Scott Laboratories, Inc.,
2955 Kerner Blvd.,
P.O. Box 9167,
San Rafael, Cal. 94912;

The Wine Lab;
1200 Oak Avenue,
St. Helena, Cal. 94574;

Tri Bio Laboratories,
1400 Fox Hill Road,
State College, Pa. 16901;

Vinquiry,
301-D East Street,
Healdsburg, Cal. 95448.

Suppliers of Chemical Equipment and Chemicals

Many winemakers will find that chemical equipment and chemicals can be useful in both wine analysis and wine production. There are many chemical supply houses. The following three offer equipment and chemicals and have regional 'distribution centers in many cities: Fisher Scientific, 711 Forbes Blvd., Pittsburg, Pa. 15219; Sargent-Welch, 7300 North Linden Avenue, Skokie, Ill. 60076; and VWR Scientific, P.O. Box 3200, San Francisco, Cal. 94119.

Larger chemical supply houses sell top-quality measuring instruments such as balances, refractometers, pH meters, colorimeters, and conductivity meters at top prices. Small winemakers interested in less expensive equipment may find that some of the newer small suppliers offer good value in less fancy equipment of this type. The following list of small supply houses is incomplete but representative:

Cole-Parmer Instrument Company,
7425 North Oak Park Avenue,
Chicago, Ill. 60648;

Extech International Corporation,
114 State Street,
Boston, Mass. 02109;

Hach Chemical Company,
P.O. Box 389,
Loveland, Col. 80537;

Lazar Research Labs, Inc.,
920 N. Formosa Avenue,
Los Angeles, Cal. 90046;

Markson Science, Inc.,
Box 767,
Del Mar, Cal. 92014;

Presto-Tek Corporation,
7321 N. Figueroa Street,
Los Angeles, Cal. 90041.

The author does not specifically recommend any of these suppliers nor wants to suggest in any way that suppliers not listed are less desirable. The list is provided to give readers some idea of where to obtain the many items mentioned in the text.

Additional Sources of Information

A number of periodicals provide information for grape growers and winemakers. The first one listed is the journal of the American Society of Enologists, a professional society. Its articles are scientific papers and generally require a technical background to appreciate. The other publications serve professional grape growers and winemakers but have articles written in a more popular style, accessible to most serious amateur winemakers.

American Journal of Enology and Viticulture, American Society of Enologists, P.O. Box 411, Davis, Cal. 95617;

American Wine Society Journal, American Wine Society, 3006 Latta Road, Rochester, N.Y. 14612;

Eastern Grape Grower & Winery News, Box 329, Watkins Glen, N.Y. 14891;

Practical Winery, 15 Grande Paseo, San Rafael, Cal. 94903;

Wines & Vines, 1800 Lincoln Avenue, San Rafael, Cal. 94901.

There are also periodicals aimed at amateur winemakers and suppliers, and most amateur winemaking supply shops carry one or more. The unofficial organ of the Home Wine and Beer Trade Association is *Beverage Communicator,* Box 43, Hartsdale, N.Y. 10530.

Grape-growing and winemaking books in print can be ordered

through most book stores. The following book sellers handle some current wine books but specialize in rare and out of print books. All have catalogs:

Marian L. Gore, Box 433, San Gabriel, Cal. 91775;

Wine and Food Library, 1207 W. Madison, Ann Arbor, Mich. 48103;

Barbara L. Feret, 136 Crescent St., Northampton, Mass. 01060.

The following two university libraries have large collections of grape-growing and winemaking books and periodicals. In some cases it may be possible to borrow books from them via interlibrary loan. They may also be able to provide photocopies of specific articles for a moderate cost. It is wisest to contact a local librarian to find the best way to proceed.

University of California, Davis, General Library, Davis, Cal. 95616;

Cornell University, New York State Agricultural Experiment Station Library, Geneva, N.Y. 14456.

SUGGESTIONS FOR
FURTHER READING

History of Winemaking

H. Warner Allen, *A History of Wine,* London: Faber and Faber, 1961.

William Younger, *Gods, Men, and Wine,* London: The Wine and Food Society, 1966.

Richard B. Lamb and Ernest G. Mittleberger, *In Celebration of Wine and Life,* San Francisco: The Wine Appreciation Guild, 1980.

Leo A. Loubere, *The Red and the White: A History of Wine in France and Italy in the Nineteenth Century,* Albany: State University of New York Press, 1978.

George Ordish, *The Great Wine Blight,* London: J. M. Dent and Sons, 1972.

Maynard A. Amerine, editor, *Wine Production Technology in the United States,* Washington, D.C.: American Chemical Society, ACS Symposium Series 145, 1981, Chapter 1.

Philip M. Wagner, *Grapes into Wine,* New York: Knopf, 1976, Part I.

Leon D. Adams, *The Wines of America,* 3d ed., New York: McGraw Hill, 1984.

Hudson Cattell and Lee Stauffer Miller, *The Wines of the East,* 3 vols. (*The Hybrids,* 1978; *The Vinifera,* 1979; *Native American Grapes,* 1980), L&H Photojournalism, (620 N. Pine St., Lancaster, Pa.).

Modern Winemaking

Books are listed in decreasing order of technical sophistication.

M. A. Amerine, H. W. Berg, R. E. Kunkee, C. S. Ough, V. I. Singleton, and A. D. Webb, *The Technology of Wine Making,* 4th ed., Westport, Conn.: AVI, 1980.

M. A. Amerine and M. A. Joslyn, *Table Wines: The Technology of Their Production,* 2d ed., Berkeley: University of California Press, 1970.

A. Dinsmoor Webb, editor, *Chemistry of Winemaking,* Washington, D.C.: American Chemical Society, Advances in Chemistry Series 137, 1974.

Maynard, A. Amerine, editor, *Wine Production Technology in the United States,* Washington, D.C.: American Chemical Society, ACS Symposium Series 145, 1981.

A. H. Rose, editor, *Alcoholic Beverages,* New York: Academic Press, 1977.

A. Massel, *Applied Wine Chemistry and Technology,* London: Heidelberg Publishers, 1969.

Richard P. Vine, *Commercial Winemaking: Processing and Controls,* Westport, Conn.: AVI, 1981.

Philip M. Wagner, *Grapes into Wine,* New York: Knopf, 1976.

Desmond Lund, *Leisure Winemaking,* Calgary, Canada: Detselig Enterprises, 1978.

Robert Benson, *Great Winemakers of California,* Santa Barbara, Cal.: Capra Press, 1977.

Maynard A. Amerine and Vernon L. Singleton, *Wine: An Introduction,* 2d ed., Berkeley: University of California Press, 1977. An excellent overview.

Harold J. Grossman, *Grossman's Guide to Wines, Beers, and Spirits,* 7th ed., revised by Harriet Lembeck, New York: Scribner's, 1983. Contains a very broad look at wines, including presentation, sales, storage, regulations, taxes, statistical data, and so forth.

Grape Growing

John R. McGrew, *Guide to Winegrape Growing,* American Wine Society (3006 Latta Road, Rochester, N.Y. 14612), 1980. For the home grape grower who plans a fairly small vineyard. Lists many sources of information and supplies.

Philip M. Wagner, *A Wine-Grower's Guide,* 2d ed., New York: Knopf, 1978. Emphasizes grape growing outside of California.

Robert J. Weaver, *Grape Growing,* New York: Wiley, 1976. Applies mainly to the hotter areas of California.

A. J. Winkler, James A. Cook, W. M. Kliewer, and Lloyd A. Lider, *General Viticulture,* Berkeley: University of California Press, 1974. A university textbook filled with detail on commercial grape production in California.

D. P. Pongracz, *Practical Viticulture,* Cape Town, South Africa: David Philip, 1978. A European viewpoint.

AGING WINE IN WOODEN BARRELS

R. G. Peterson, "Research Note: Formation of Reduced Pressure in Barrels During Wine Aging," *Am. J. Enol. Viticult., 27,* 80 (1976).

V. L. Singleton, "Some Aspects of the Wooden Container as a Factor in Wine Maturation," Chapter 12, *Chemistry of Winemaking,* Washington, D.C.: American Chemical Society, Advances in Chemistry Series 137, 1974.

FORTIFIED WINES

M. A. Joslyn and M. A. Amerine, *Dessert, Appetizer & Related Flavored Wines: The Technology of Their Production,* Berkeley: University of California, 1964.

Wyndham Fletcher, *Port: An Introduction to Its History and Delights,* London: Sotheby Parke Bernet, 1978.

Fortified wines are also discussed in Amerine et al., *The Technology of Wine Making;* Webb, ed., *Chemistry of Winemaking;* and Rose, ed., *Alcoholic Beverages.* (See above.)

NONGRAPE WINES

Homer Hardwick, *Winemaking at Home,* Scranton, Penn.: Funk and Wagnalls, 1970.

Stanley Anderson and Raymond Hull, *The Art of Making Wine,* New York: Hawthorn, 1971.

Roger A. Morse, *Making Mead (Honey Wine),* Ithaca, N.Y.: Wicwas Press, 1980.

John Ehle, *The Cheeses and Wines of England and France with Notes on Irish Whiskey,* New York: Harper & Row, 1972.

See also Rose, ed., *Alcoholic Beverages* (above).

WINE ACIDITY

M. A. Amerine, R. M. Pangborn, and E. B. Roessler, *Principles of Sensory Evaluation of Food,* New York: Academic Press, 1965, pp. 75–82.

R. A. Plane, L. R. Mattick, and L. D. Weirs, "An Acidity Index for the Taste of Wines," *American Journal of Enology and Viticulture, 31,* 265 (1980).

L. R. Mattick, R. A. Plane, and L. D. Weirs, "Lowering Wine Acidity with Carbonates," *American Journal of Enology and Viticulture, 31,* 350 (1980).

H. W. Berg and R. M. Keefer, "Analytical Determination of Tartrate Stability in Wine. I. Potassium Bitartrate," *American Journal of Enology and Viticulture, 9,* 180–193 (1958).

LABORATORY ANALYSIS OF WINES

M. A. Amerine and C. S. Ough, *Methods for Analysis of Musts and Wines,* New York: Wiley, 1980.

SENSORY EVALUATION OF WINES

M. A. Amerine and E. B. Roessler, *Wines: Their Sensory Evaluation,* San Francisco: Freeman, 1983. Very scientific in approach, with details on professional-level wine judging.

S. W. Andrews, *Be a Wine and Beer Judge,* Andover, England: Amateur Winemaking Publications, 1977. Contains details on well-run amateur judgings.

Michael Broadbent, *Wine Tasting,* 6th edition, London: Christie's Wine Publications, 1979. For the average wine consumer.

For background on sensory evaluation in general, see Amerine et al., *Principles of Sensory Evaluation of Food* (above).

WINES IN COOKING AND MIXED DRINKS

Anne M. Logan, *Wine and Wine Cooking,* Richmond, Va.: Media General Publication, 1972.

Bob Sennett, editor, *Complete World Bartender Guide,* New York: Bantam Books, 1977.

VINEGAR MAKING

M. A. Amerine, H. W. Berg, R. E. Kunkee, C. S. Ough, V. L. Singleton, and A. D. Webb, *The Technology of Winemaking,* 4th ed., Westport, Conn.: AVI, 1980, pp. 557–559 and 650–652.

Modane Marchbanks, "Vinegars," in *The Family Creative Workshop,* New York: Plenary Publications, 1976, Vol. 22, pp. 2694–2703.

INDEX

Library of Congress Cataloging in Publication Data

Jackisch, Philip, 1935–
   Modern winemaking.

   Bibliography: p.
   Includes index.
   1. Wine and wine making—Amateurs' manuals.
I. Title.
TP548.2.J33   1985   663'.2   84-45803
ISBN 0-8014-1455-5 (alk. paper)